WINGS ACROSS THE BORDER

A History of Aviation in North East Wales and the Northern Marches

Volume I

Derrick Pratt & Mike Grant

bridge
books

Wings Across the Border
First published in Wales by
BRIDGE BOOKS
61 Park Avenue
Wrexham, LL12 7AW

To

Veronica and Shirley

A CIP entry for this book is available from the British Library

ISBN 1-872424-69-4

Printed and bound by
Design 2 Print
Llandudno

Contents

Abbreviations

A	Acting
AACU	Anti-Aircraft Co-operation Unit
AAEE	Armament & Aeroplane Experimental Establishment
AAP	Aeroplane Acceptance Park
AB	Able Seaman
AFC	Australian Flying Corps
AFV(D)	Armoured Fighting Vehicle (Depôt)
AGS	Air Gunnery School
Airco	Aircraft Manufacturing Co.
AM1, 2, 3	Air Mechanic (1st, 2nd 3rd Class)
AOC	Air Officer Commanding
ARS	Aircraft Repair Section
ASU	Aeroplane Storage Unit
ATC	Avro Transport Company
BE	Blériot Experimental
BOAC	British Overseas Airways Corporation
C	Coastal (airship type)
C	Armed reconnaissance plane (German classification)
Cdt.	Cadet
CEF	Canadian expeditionary Force
CFS	Central Flying School
CO	Commanding Officer
Cpl.	Corporal
CWGC	Commonwealth War Graves Commission
D	Single seater fighter (German classification)
DFW	Deutsche Flugzeug-Werke
DA&AAP	Directorate of Aerodromes & Air Acceptance Parks
DH	de Havilland
DSO	Distinguished Service Order
E/F	Engine failure
F/Cdr.	Flight Commander
F/Cdt.	Flight Cadet
FE	Farman Experimental
F/Lt.	Flight Lieutenant
FSL	Flight Sub-Lieutenant
FTS	Flying Training School
GAC	Gloster Aviation Company
GS	General Service
GWR	Great Western Railway
HE	High Explosive
HMA	His Majesty's Airship
HP	Handley Page
h.p.	horsepower
IWGC	Imperial War Graves Commission
Jasta	Jagdstaffel (German scout unit)
KSLI	King's Shropshire Light Infantry
L	Luftschiff (German airship numbering)
LAM	Leading Air Mechanic
Lt.	Lieutenant
Ltn.	Leutnant
LNWR	London&North Western Railway
MC	Military Cross
MF	Maurice Farman
Mk.	Mark
ML	Motor Launch
MSFU	Merchant Ship Fighting Unit
MT	Mechanical Transport
MU	Maintenance Unit
NAPS	Naval Airship Patrol Station
NAF	National Aircraft Factory
NAVEX	Navigational Exercise
NCO	Non-Commissioned Officer
NT	Norman Thompson
Obs.	Observer
OC	Officer Commanding
OMFC	Overseas Military Forces (Canadian)
OR	Other Rank(s)
ORB	Operations Record book
OS	Ordnance Survey
OSL	Observer Sub-Lieutenant
OSRAP	Observer School of Reconnaissance and Aerial Photography
(P)AFU	(Pilot) Advanced Flying Unit
P/O	Pilot Officer
POW	Prisoner of War
PPCLI	Princess Patricia's Canadian Light Infantry
Pte. 1, 2	Private (1st, 2nd Class)
Q, QFQ	(decoy) site; night-time decoy
R	Rigid (airship numbering)
RAeC	Royal Aero Club
RAE	Royal Aircraft Establishment
RAF	Royal Aircraft Factory (pre–April 1918)
RAF	Royal Air Force (post–April 1918)
RCAF	Royal Canadian Air Force
RFC	Royal Flying Corps
RLG	Relief Landing Ground
RNAS	Royal Naval Air Service
RNAS	Royal Naval Air Station
RNVR	Royal Navy Volunteer Reserve
ROC	Royal Observer Corps
ROF	Royal Ordnance Factory
RS	Reserve Squadron
RWF	Royal Welsh Fusiliers
S/Cdr.	Squadron Commander
SDF	Special Duties Flight
SDS	Special Duties Station
SE	Santos Experimental
S/Ldr.	Squadron Leader
SNO	Senior Naval Officer
Sqn	Squadron
SS	Sea Scout (airship)
SSE	Sea Scout Experimental
SSP	Sea Scout Pusher
SSZ	Sea Scout Zero
STS	Seaplane Training School
T	Temporary (with ranks)
TA	Territorial Army
TDS	Training Depôt Station
TS	Training Squadron/School
USAS	United States Air Service
WAAC	Women's Army Auxiliary Corps
W/Cdr.	Wing Commander
WO	Warrant Officer
WRAF	Women's RAF
W/T	Wireless Telegraphy (equipment/operator)

Introduction

This is the first volume of a planned series covering aspects of aviation history in north-east Wales and border region. Chronologically it explores the period 1780–1930. Although the narrative regularly zooms in upon the parish pump, this is not a 'local' history in the popularly accepted sense of the term. It is more of a macro-history covering an area extending from Anglesey, along the North Wales holiday strip and Dee estuary, south to include Chester, Wrexham and the middle Dee valley before petering out in the northern March somewhere south of Shrewsbury. In other words the authors treat of the lowland areas skirting the Welsh mountain massif, underscoring spheres of interest and lines of movement dictated over the centuries by the vagaries of Welsh geography. It is an area in which they are most immediately at home and on which they can write with some authority.

If there has to be a focus it is Wrexham, no less than any other border town a venue for the earliest demonstration flights, barnstormers and flying circuses. It still boasts the scant remains of a Second World War aerodrome surprisingly unsung and unhonoured by local historians and aviation buffs alike, but which had a surprisingly long peace time pedigree and gestation period. But while to treat of an individual airfield, bombing range, decoy site and other military installations is indeed 'local' history, focussing on a specific point, the aircraft that flew from them, used them, or were, hopefully, deceived by them, covered great distances in a relatively short space of time. To write local history on an aviation theme, therefore, parish parameters have of necessity to be extended, very much in the same way as local historians handle other 'linear' themes such as the history of canals, roads and railways.

The authors are both natives of Wrexham. Derrick Pratt is a professional geographer and for twenty-five years was part-time tutor in local history and landscape studies in the Extra-Mural Department of UCW, Bangor. Ruefully he is the first to admit that, although intrusive elements into the landscape, airfields, bombing and gunnery ranges, MUs, radar stations, decoy sites etc. received but cursory treatment and were peripheral to archaeological and medieval studies. However, eight years ago interest and research were directed into new channels whilst attempting to unravel the events of the night of 6 May 1944. A de Havilland Mosquito II, serial DZ747, of 60 OTU High Ercall, buried itself almost unnoticed in the soft made-up ground that had been the 'Outer Moat' of Penley Hall, perhaps better known then as No. 129 American General Military Hospital, just across the road from the Maelor School where he had taught for thirty-one years. The Mosquito had been on a night NAVEX over the Irish Sea. When control was lost in bad weather, the pilot, F/Lt. J. R. Milne, stayed with his plane and was killed. His navigator, P/O Rice, managed to bale out, landing in Park Lane, Penley, where he was promptly apprehended as an enemy parachutist!

Mike Grant's lifetime interest in aviation began when, as a small boy, he was allowed to scramble around the Airspeed Oxford exhibited in the Parciau, Wrexham, as part of the town's VE-Day celebrations in 1945. Through the good offices of the curator of the Aerospace Museum, Cosford, Derrick was put in touch with Mike. A trained mechanic, the latter had been recruited by F/Lt. J. F. Worth as a civilian member of the Aerospace Society preparing aviation heritage material for display at the RAF Museum. It was was but a short step to membership of the Wartime Aircraft Recovery Group (WARG), originally based at Cosford but now housed on the former aerodrome at High Ercall. After working on the restoration of engines and airframes, Mike was appointed the Group's archivist and pre-recovery crash-site researcher, a natural progression, perhaps, for one who is also a member of the RAF Historical Society, an Air Britain historian and a member of the Cross & Cockade International Society of World War 1 Aviation Historians. Without his detailed knowledge of the aviation records at, and the mysterious workings of, the PRO, the abstraction of the raw material for this series of books would not have been accomplished so smoothly and efficiently.

As the two Wrexham researchers got together they quickly realised (a) how little existed in the way of a systematic study of the aviation heritage of Wrexham and its extended hinterland and (b) how very rapidly much of that story was being irretrievably lost, both in the destruction of tangible physical remains and the inevitable disappearance of valued eye-witnesses who could supply the stuff of oral history.

Eight years of fruitful co-operation followed, of which this present volume, the first in a series, is the tangible by-product as an inquisitive local historian turned his attention to a new exciting world of balloons, airships and aeroplanes. There is more to aviation in north-east Wales and border country than hitherto suspected. In the First World War RFC aerodromes sprang up, rather belatedly, at Sealand, Shawbury, Monkmoor, Tern Hill and Hooton Park. RNAS stations were established at Llangefni and Bangor. Inevitably their existence was marked by numerous flying accidents and fatal crashes which had an enduring impact on rural life and country churchyards.

Twenty years on, in a second world conflict, Merseyside's war was also Wrexham's war. Wrexham was on 'Adolf's Railway', favoured Luftwaffe route up the Welsh border following the Severn and Dee, turning right at Shocklach for Manchester and Crewe. RAF Wrexham was the only 9 Group night-fighter station in the north-east borderland purposely built to deny this back door to the Luftwaffe. Other airfields (many with familiar names) — Hawarden, Sealand, Tilstock, Tern Hill, Shawbury, Sleap, Hooton Park, Cranage, Montford Bridge, Rednal, Calveley, Condover and their satellites — were essentially training aerodromes that occasionally had defensive roles thrust upon them willy-nilly as the need arose. Their histories will be touched upon in successive volumes on this present series.

Not surprisingly, during daylight hours, middle Dee air-space was densely crowded with learner aircraft, which makes for some dismal statistics. In Flintshire, Denbighshire and Cheshire some 1,000 aircraft suffered accidents or were destroyed on and off airfields between 1917–45, along with six German machines. In Shropshire, the figure rises somewhat dramatically to over 1, 800 Allied machines and three German aircraft lost or damaged. Few places were left unscathed by 'fall out' of such dimensions. Two air wars did indeed have a considerable impact one way or the other on the every-day life and landscape of the area.

But the historian also discovers that the north-east borderland had not been left unscathed by the progress of aviation in the years immediately preceding the First World War and the two decades of peace that followed it. Gustav Hamel's proselytising visits; Ernest Maund, would-be aircraft constructor of Church Stretton; Vivian Hewitt, Rhyl's aviator extra-ordinary; the infatuation of Thomas Murthwaite Dutton, Queensferry garage proprietor, with engines and aircraft; the passage of Anglesey blimps and the giant rigid airships, the R33, R101 and Graf Zeppelin; a Shropshire hero whose death in 1912 marked the premature abandonment of the monoplane by the military — such are only some of the many isolated facets which, when brought together, provide a fascinating chronological insight to local history from an unusual aspect, the air. This is not to forget that before the aeroplane there were the balloons, flown by daring, even foolhardy, aeronauts from towns along the Welsh border, culminating in the annual two-day onslaught on Welsh air-space by the better handled 80,000cu. ft contraptions that for thirty-four years (1880–1913) were the main attraction at the Shrewsbury Flower Shows.

Aviation in this part of the world is not dead. The culmination and latest manifestation of almost a century of aviation in North Wales is the daily passage of a giant Beluga aircraft making its round trip between Toulouse and the British Aerospace Airbus production facility at Broughton, Flintshire. The latter has evolved out of the old Vickers Armstrong shadow aircraft factory of 1937 vintage. The runway is an updated, extended hi-tech version of that which once echoed to the roar of 57 OTU's Spitfires and 41 OTU's Mustangs and Hurricanes at RAF Hawarden. Thomas Murthwaite Dutton, Flintshire's own aviation pioneer, was born at Beech House just over the boundary fence, in turn but a few yards from Cop House Farm the site of Sir Alan Cobham's barnstorming operations. So an almost unbroken link is forged. One wonders what Dutton would have made of all this?

In putting together this story the authors are conscious of (a) the fact that their research is still on-going, but a line has to be drawn somewhere, and (b) the great debt owed to several hundred individuals—a vast, but sadly diminishing, source of oral history — and major record repositories. Amongst the latter special thanks are due to archivists, librarians and staff at the Anglesey, Caernarfonshire, Denbighshire, Flintshire, Chester and Cheshire Record Offices, the Shropshire Records and Research Centre, and Wrexham, Llangollen, Mold, Rhyl, Oswestry, Whitchurch, Shrewsbury, Market Drayton and Darlaston libraries.

For this particular volume thanks are due to: staff of the PRO, Kew; the Air Historical Branch, Air Ministry; Air Britain Historians (especially Peter Moss); RAF Museum, Hendon; Aerospace Museum, Cosford (L. Woodgate); Commonwealth War Graves Commission (Maria Ghoules); members of the RAF Historical Society and Cross & Cockade International; British Telecom plc Archives (Yvonne Smith); Imperial War Museum (Dept. of Printed Books); Institution of Electrical Engineers (T. Proctor); National Library of Wales, Aberystwyth; National Maritime Museum, Greenwich; Science Museum (C. N. Bunyan, Jane Insley); Fleet Air Arm Museum, Yeovilton; Airship Heritage Trust, Cardington (especially the magazine *Dirigible*); National Motor Museum, Beaulieu (Michael Budd); Patent Office, Newport, Gwent (M. K. Rees).

Many individuals have been interviewed or have supplied information and photographs. Special thanks are due to: D. J. Barnes, Bury; C. N. Barker, Hawarden; Leslie Bellis, Wrexham; Mrs. Bennett, Frog Hall; Harry Brereton, Tilston; J. M. Bruce, author; Capt. M. J. P. Cowper, RAF; Colin Crowley, Llay; Colin Cruddas; Mr. Davies, Sealand Road, Sandycroft; Ken Davies, Northop; Raymond Davies, Mold; K. H. Day, RAF/ATA; Alton Douglas, Birmingham; W. M. Edwards, Wrexham; M. Evans, Borras; A. E. Harrison, Hawarden; Doreth Hughes (ex-Shotwick WAAC); David Hughes, Caergwrle; the late Mrs. M. Grant (of Rubery Owen, Whitegate); George Hughes, Shocklach; Col J. Margerson, Queensferry; Messrs Mason & Moore Dutton, solicitors, Malpas and Chester; Bernard Mitchell, Llanidloes; Derek Parker, Holt; Lt/Col J. Starling, RFC; the late F/Lt. G. R. Sunderland, RAF (retd.); Mrs. C. Temple, Wrexham; Simon Warburton, Grosvenor Museum, Chester; F. Wilton, Sealand; Ernest Woodfine, Isycoed.

The authors have made every effort to trace the original owners of photographs included in this book, but many have come sixth or seventh hand obviously from personal photo albums of service personnel long deceased; others have gained wide currency amongst aviation enthusiasts through repeated copying and it is difficult to trace originals; all have been collected rather haphazardly over some forty years, which tends to further obfuscate matters of ownership. Apologies, therefore, for the quality of some of them. Sources, where known, have been indicated.

Balloon fever

Between Cefn Park and Llwyn Knottia farm on the eastern fringes of Wrexham, above what used to be a boating lake, the 200ft contour defines an elongated, gently swelling ridge known as *Balloon Hill*. In the aviation world the wooded western end of the ridge is known only as the crash site of the 'Wrexham Heinkel', a German bomber brought down by a Defiant of 96 Squadron on the night of 7/8 May 1941.

For some local historians, however, the very name *Balloon Hill* would serve to extend the history of flying in the Wrexham area back some 150 years, at least half a century before Gustav Hamel in his Blériot monoplane first dazzled the crowds at the Racecourse. Unfortunately this particular hypothesis remains unsubstantiated by documentary evidence or contemporary references in local newspapers The name itself would appear to be relatively late, appearing for the first time on the Ordnance Survey 6-inch map second revision of 1910 (so-called Third Edition). It is absent from the 1872 6-inch map, and in the Erlas Tithe Apportionment of 1842 the hill is known simply as Cottage Field. Perhaps the safest course is to accept that it is purely topographical, from the close resemblance of the ridge to a semi-inflated dirigible. Yet there will always remain the nagging doubt that sometime during the hey-day of ballooning an aerosat did make an unheralded landing here, blown across the Dee from Cheshire by unfavourable winds.

A parallel might be drawn with *Aeroplane Field*, a name given by an older generation of Wrexham inhabitants to several fields on the edge of town — at the Hightown Barracks and in Cefn Road, Rhosnesni, for example — that reflect their regular, or even 'one-off', use by military and pre-War civilian private aircraft. *Aeroplane Field* is always used in speech, never in writing. Within another generation this random renaming of a particular field will fade into oblivion as popular etymology wroughts its transforming work again and substitutes yet another familiar, and perhaps more appropriate, field-name.

It is an unpalatable fact that, apart from the attempts (1812–17) of James Sadler and his sons to make the Dublin–Holyhead crossing of the Irish Sea, and Lord Newborough's eccentric dabblings with unmanned balloons at Glynllifon in the 1850s, ballooning in North Wales and border regions was confined almost exclusively to select Chester, that provincial metropolis, ecclesiastical centre, military depot, and tourist place.

In 1799 Chester was described by its mayor as 'not a place of trade, though the residence of many opulent and independent people'. Such a city, with a population of some 15,000, was preferred by the early astronauts to insanitary Wrexham and the squalor of its satellite industrial villages.

Early balloon flights or ascents were organised on the lines of a fairground attraction. Until the novelty wore off crowds flocked to see these giant primitive aerosats launched, as much for the chance of witnessing a terrible accident as for the flight itself. The more successful pioneer balloonists toured the country, using more and more tricks and novelty items to attract paying spectators. On this last criterion alone Wrexham, with its motley population in the 1780s hovering about the 3,500 mark, did not figure in their itineraries. But the town did not remain immune to the balloon mania that swept the country in the closing decades of the 18th century. Its gentry, business people, artisans, hustlers and petty criminals made the obligatory pilgrimage to Chester's Roodee, Castle and gas works to watch and participate in balloon flights and ascents, or simply to gawp and marvel at the intricate process of inflation.

In November 1783 the first manned Montgolfier (hot air) balloon flight took place near Paris, followed two weeks later by a Charlière hydrogen filled balloon, a much more practical piece of apparatus, giving a longer sustained flight. These pioneering efforts captured the imagination of Europe. By August/September 1784 the first successful experimental flights had taken place in England. Prominent amongst the amateur aeronauts was one Vincenzo Lunardi, a servant or clerk to the Neopolitan Ambassador to the Court of St. James, who flew his Charlière from London's Artillery Ground on 15 September 1784. Lunardi would make twelve further flights in this country before a fatal accident to a spectator in 1786 forced his early retreat to mainland Europe.

News travelled fast, even to Wrexham. On 19 September 1784 Brownlow Cust could write to his brother-in-law, Philip Yorke, at Erddig: 'Simon Yorke and Mrs. R. Cust had had a good view of Mr. Lunardi's Balloon from the new River Head'. Today the sound of aircraft is ever in our ears. Wrexham people, living along the flight path to and from Manchester Airport and overphased by RAF Shawbury's helicopters, read only with the mildest curiosity of Thomas Birch's awe at his first sight of a manned balloon. A London lawyer and close friend of the Erddig family he wrote on 22 October 1784 to Philip Yorke, ostensibly to thank him for a present of woodcock, but could digress at length: ' I was amongst others to see Blanchard & Sheldon go up with their balloon, it was really an awefull sight to see two Fellows mount the Skies & take the wings of the wind relying on a machine, so liable to accidents, that would precipitate them to the Earth & dash them in Pieces, to See them mount, bowing & kissing their hands with as much ease, as if only getting into a Phaeton, is truly surprising, & shews the curious & daringly adventurous Spirit of Man. I thank God they are both safe on Land'. This is a reference to the rather acrimonious ascent of Jean-Pierre Blanchard and Dr. John Sheldon from Lochee's Military Academy, Chelsea, on 16 October 1784, in a boat-shaped car equipped with virtually useless hand-operated wings, rudder, and rotating fan as by now standard propelling and steering devices.

Following his London triumph Lunardi constructed a bigger and better balloon, decorating its envelope with the Union Jack. In July 1785 he embarked upon a provincial tour, making ascents at Liverpool (2), Chester, Edinburgh (2), Kelso, and Glasgow (2). His two Liverpool ascents were described by the local press as both humdrum and foolhardy. Problems with inflation forced Lunardi to take off in adverse weather conditions rather than face the anger of a fickle, frustrated crowd. The flight of 20 July, after being blown out over

Lunardi's second balloon under inflaltion. It appeared at Chester in August 1785. Note hydrogen supply, the car with oars and wings, and poles to support the semi-inflated balloon. [Science Museum]

Liverpool Bay, eventually saw him dumped unceremoniously in Simon's Wood, beyond Aintree, after being airborne for sixty-three minutes. For the ascent of 9 August dangerously strong north-westerly winds were blowing. After an hour Lunardi made a rough landing in a cornfield at Tarporley. But his grappling-hook would not hold and he suffered the most terrifying indignity facing a balloonist — being dragged willy-nilly through hedges and trees, demolishing a cottage chimney in Tiverton before finally being brought to rest, with the help of several farm labourers, in a lane near Beeston Castle.

After this adventure Lunardi brought his Charlière 'flaming air' hydrogen balloon to Chester Castle yard. There was plenty of room as Thomas Harrison's powerful Greek Revival remodelling of the castle and outer bailey would not commence until 1788. There was the added advantage of half a company of the Royal Welch Fusiliers to guard balloon and wagons and willing hands of Cheshire militiamen to man the ropes and poles that supported the envelope during inflation. However, on Monday, 22 August 1785 gusting winds hindered Lunardi's first attempted ascent. A disappointed chronicler wrote: 'It was rather an uninteresting scene owing to several accidents which occurred to the complicated apparatus then used in the process of inflation. We shall therefore pass over this and proceed to that of Lieut. French, R. C. M. on Wednesday, 30 August 1785... ' Lunardi's flying companions, a dog, cat and pigeon (not the most viable and harmonious of working groups) were spared the cold and misery of yet another ascent.

The Charlière balloon incorporated many features still in use two centuries later — a net that enclosed the gas bag, a valve to release gas, and ballast that could be dropped to reduce the rate of descent. Once airborne, the flights of the hydrogen balloon were now measured in hours instead of minutes.

It took eight days to repair the damage to the balloon envelope. It is clear from the terse record of the ascent that Lt. French made a solo flight, but one can only hazard a guess as to what prompted Lunardi to hire out his precious balloon to Chester's novice aeronauts. Normally, with Lunardi as pilot, a male passenger would provide the second pair of hands needed to man the 'wings' and two 'oars' with which, in his early balloons at any rate, Lunardi sought to control movement of the balloon in both horizontal and vertical planes. It may be that he had already decided, in view of the battering his balloon had received at Liverpool, to lay down the keel of another, of similar shape and design, which, after proving trials, he would take north with him to Edinburgh.

Balloons would be used briefly, but with considerable success, by the French for military observation purposes during the Napoleonic Wars. However, their use was discontinued on the personal order of the Emperor himself. Possibly, as an unkind commentator would later remark, 'he felt they detracted from his own military prowess'. The British Army showed little interest in them. In young Lt. French of a backwood's militia regiment we have, perhaps, that rare phenomenon, a thinking officer and strategist far ahead of his time.

The contemporary diarist continues: 'Yesterday afternoon, about four o'clock, Lieut. French of the Royal Cheshire Militia, ascended with Mr. Lunardi's balloon from the Castle yard in this city; the day was fine. During his voyage he threw down his hat and coat to prolong his continuance in the air, which was about two hours, when he descended in the greatest safety near Macclesfield in this county, distant upwards of 30 miles. . . ' Commandeering carts from an apprehensive peasantry to load balloon and gondola, Lt. French spent the night being feted at a local hostelry before returning in triumph to Chester the following day, being escorted in from the city boundary by the band of his regiment.

Nine days later there was a third balloon ascent from Chester. However, our chronicler is a little ambiguous and leads one to believe that the junketing actually involved two balloons, that of Lunardi, and a smaller balloon constructed by a certain Thomas Baldwin,

prosperous city merchant, which was inflated slightly earlier. The latter may have been used for display purposes, to keep the interest of the crowds, or as a pilot balloon released to ascertain the strength and direction of the wind — a regular feature of later launchings. Mr. Baldwin had very early been bitten by the ballooning bug, for his aerosat had apparently been completed some two years previously and was just awaiting a propitious moment for launching. Whatever, even as rich adventurers of a later period would turn to motor racing or the aeroplane, so the Chester merchant could afford to hire Lunardi's balloon, and on 8 September 1785, with the great aeronaut himself supervising the complicated inflation process, Thomas Baldwin's dreams turned into reality:

'On the 7th instant was the day appointed for [the] aerial excursion of Thomas Baldwin, Esq., a native of this city, from the Castle Yard; which was postponed to the following day, on account of the violence and unsteadiness of the wind; Mr. Lunardi declaring that he would not attempt to inflate his Balloon until the weather had abated. On Thursday the morning was bright and calm below, with an upper current of light clouds moving from the S. W. About nine o'clock a cannon was fired to inform the neighbourhood. At a quarter past ten, an aerostatic globe, 18ft in circumference, contrived and completed by Mr. Baldwin near two years ago, [began to take shape].

'At twelve o'clock the second cannon was discharged to notice that the balloon was half inflated; and at half past one our intrepid aeronaut stepped into the car amidst the acclamation of his fellow citizens, and applause that was mixed with terror and delight. By his assistance, Mr. Lunardi's balloon was inflated and with much less trouble and expense than before.

'The balloon rose perpendicularly with great velocity, and in the grandest manner. Mr. Baldwin, it is not doubted but the public will receive both information and amusement [sic]. We shall only add that our aeronaut descended at Kingsley, ten miles from Chester [north of Delamere Forest], re-ascended at much greater height than at first, and finally descended five miles from Warrington. The following day he returned and was met by the militia band and ushered with loud 'Huzzas!' into his native city. His flags were carried in procession, the bells rang, and every other demonstration of joy was shown on the occasion'.

These ascents underscore the fact that, although in theory an accomplished aeronaut would ascend and descend to make maximum use of favourable air currents at various altitudes, in practice these early pioneers were entirely at the mercy of the wind. In this respect the choice of Chester for their purpose may seem strange unless the rattle of money at the tills over-rode any other consideration. Promoters could boast about unrivalled prospects of the Welsh mountains and the estuaries of the Dee and Mersey, but in the last resort, with wayward winds, these could prove dangerous to pilot and passengers. It is only good fortune that Chester based balloonists have not drifted out over the Irish Sea, or, worse still, the Atlantic! Hence the unlikely selection in 1862 of the Wolverhampton Stafford Road gas works for an attempt on the then world altitude record!

Following the departure of Vincenzo Lunardi from England in 1786, Chester did not witness another major balloon event until 1824, with the ascent of a 'Mr. Sadler'. This gentleman was either John or Windham Sadler, probably the latter, aeronaut sons of James Sadler, who, in October 1812, almost accomplished the elusive Dublin–Anglesey crossing of the Irish Sea, to be foiled by an uncharacteristic bout of overweening ambition and contrary winds that dumped him unceremoniously in Liverpool Bay!

Sadler passed his ballooning skills on to both his sons, but it was Windham who successfully made the first Irish Sea crossing in July 1817. The latter also experimented with a new lifting agent, coal gas, which was becoming more readily available as towns built gas works for their street and domestic lighting schemes. The first coal gas ascent

James Sadler's ascent from Dublin in October 1812. Owing to contrary winds he missed Anglesey and made a forced landing in Liverpool Bay. [Science Museum]

was made on 19 July 1821 by the noted aeronaut and balloon manufacturer Charles Green, who would make 526 balloon ascents in a career spanning some thirty-one years.

The contemporary account of Mr. Sadler's 'Grand Ascent' from the Castle Yard in June 1824 conveys something of the razzamatazz that surrounded such events. The novelty aspect still drew the crowds but the steps taken to relieve them of their money are even more in evidence: 'In order to exclude those who did [not] pay witnessing the ascent, the open spaces on the right and left of the Great Entrance were screened by sail cloths and tarpaulins affixed to poles, all of which were gratuitously furnished by our respected townsman, Thomas Dixon, Esq. The preparations were completed on Saturday night last, and on Sunday the Castle Yard was visited by some thousands from the city and neighbour[hood].

'Monday came, and a more glorious morning was never witness[ed]. The hour fixed for the ascent of the balloon was two o'clock, but at [that] time it was little more than half filled. Three, four and five o'clock arrived. At half past five o'clock another cannon was fired and the car was brought out. All was now bustle and excitement. There was a general rush in which the Balloon was placed; and the guard of the 53rd Regiment on duty had difficulty in repressing the swell of the curious crowd. Mr. Sadler now appeared in his travelling dress, a blue jacket and trousers, and a seal's-skin cap, and the gas tube was detached from the balloon and firmly closed. The car had previously been attached and Mr. Sadler got into the car to try the buoyancy of his magnificent vehicle.

This part of the ceremony was performed by Lady Standley (who approached under Sir Thomas Standley's arm) and Mrs. Barnston. A few appropriate words were addressed to Mr. S., who made a suitable reply. Taking the colours into the car the adventurous traveller called out 'Let go!' and the balloon sprang from the grasp of those who held it. The spectacle was truly sublime!

'The gorgeous machine slowly ascended into the air amidst the shouts of the admiring multitude and the thunder of cannon. It passed

majestically over Julius tower [the Julius Agricola Tower of the inner bailey, the principal medieval survival], and took a direction nearly S.S.E. Mr. S. waved his flags and [acknowledged] the cheers of the spectators by taking off his cap. He continued gradually ascending and at the altitude of some 300 or 400 yards, he threw out a number of cards, the descent of which had [the] appearance of so many pigeons skimming the air in different directions and afforded a very pleasing sight. Ten minutes after his ascent, when nearly over Eaton Hall, he was driven by a new current of wind in a direction nearly due North, and continued rapidly ascending when the wind again changed and hurried him due East, apparently in line for Beeston Castle. At this time, Mr. S. informs us, he was nearly three miles in height, and the scene in his progress was of the superb description. Chester Castle was in view till a few minutes before his descent. Looking northwards the scene extended to the Ribble, Bootle Bay, Mersey and Dee. He distinctly saw the steam boats plying between Birkenhead, Tranmere and Liverpool, whilst beneath him was the Vale Royal of England, and Congleton, Macclesfield and the Hills of Derbyshire in the perspective view.

'When the balloon ascended it was little more than two-thirds filled. The power of the sun, however, began to decline, and as Mr. Sadler had let off a great quantity of gas, he thought it prudent to prepare for his descent. Accordingly, a little beyond Tarvin, he occasionally opened the valve, and soon approximated the earth. He saw the country people running from all quarters towards Utkinton. He was caught hold of by a man and the balloon gradually descended to the ground at 19 minutes past seven o'clock without damage being sustained. Every assistance was afforded to Mr. Sadler. His balloon was placed in the car and carried to 'The Swan' in Tarporley, to which place Mr. Sadler was also carried on the shoulders of the peasantry, who cheered the bold adventurer. Many hundreds were soon collected in front of the inn and the church bells rang a merry peal.

'Very fortunately a landau in which was Mr. Gunning and two ladies from Liverpool followed the direction of the balloon and came up with Mr. Sadler soon after his descent, as did a chaise, in which [were] Mr. Tomlinson and Mr. Henshall, immediately afterwards. Mr. Sadler returned to Chester in the landau about a quarter before 11 o'clock, the flags displayed from the sides, followed by the chaise with the balloon on top of it. Crowds followed the carriages through the streets, and on alighting Mr. Sadler addressed some thousands, thanking them for their kindness and expressing regret that the ascent had been so long delayed. He then entered the hotel amidst general cheering from those without, and from many friends within, who had assembled here to greet his safe return!'

This was one of the last flights made by Windham Sadler. On 29 September 1824 he was killed whilst making an ascent at Bolton. His father (d. 26 March 1828) survived him, the only one of the pioneer aeronauts to live on into the age of the professional balloonist and great showmen. This courageous group, putting behind them the flapping wings and oars of Lunardi and Blanchard, accepted the limitations of the free balloon, mastered and exploited its more spectacular features, and earned a comfortable living thereby. No fete or celebration was complete without a balloon ascent, but out of necessity it would be from very unromantic surroundings, such as the gasworks' yards of Shrewsbury, Chester and Welshpool, that the new breed of aeronauts would rise majestically in their great colourful balloons. They numbered amongst their body both stars and clown princes!

The same time as Windham Sadler was busy in Chester the Green family arrived in Shrewsbury. They had ballooning in the blood, so much so that, when the Roll of the Aeronauts was first published in 1829, Charles Green, his brother James, sons and nephew, had amassed the astonishing total of 535 noteworthy ascents between them. Charles Green pioneered the use of coal-gas in balloons. In all his ascents he used hydrogen only once. He made his first flight on 19 July 1821 in celebration of George IV's coronation. He ascended in a magnificent red, blue and green machine of some 16,000 cu. ft. capacity, decorated with the royal arms and the inscription 'George IV Royal Coronation Balloon'. It was this balloon, now a veteran of nineteen outings, that he brought to Shrewsbury in August 1824 to make his twentieth ascent. Despite the variable quality of coal-gas nationwide, the need for larger balloons to match the lifting power of hydrogen, and being frequently ripped off by unscrupulous gas companies, coal-gas was still much cheaper. More important from the safety aspect, because of ease and speed of inflation using coal-gas, there were now fewer disastrous launchings through aeronauts deciding to face the elements rather than the wrath of a dissatisfied crowd.

From Monday, 16 August, Green's partially inflated balloon was exhibited daily at the Circus, visited by thousands who dutifully coughed up their 1s. (Ladies & Gentlemen) or 6d. (Servants & Children). As will be seen later, this tight financial control of every aspect of count-down contrasted markedly with the almost sloppy organisation of the Grahams and made the difference between being well-off and living on the breadline, haunted by creditors. Early on Monday, 23 August, the balloon was removed to Castlefields Gasworks yard, where a convenient 20-foot deep pit, excavated for a second gasometer, facilitated inflation and screening from non-paying eyes.

Zero hour was fixed for 4pm, but the inner enclosure remained only

Charles Green brings his 'George IV Royal Coronation Balloon' to Shrewsbury in August 1824. [SRRC]

Under the Patronage of the Right Worshipful the Mayor, and a Committee of Gentlemen.

MR. GREEN

Respectfully announces to the Inhabitants of SHREWSBURY and its Vicinity, that the

EXHIBITION of his MAGNIFICENT

BALLOON,

Inflated with Atmospheric Air,

Together with the CAR and its Appendages,

WILL COMMENCE ON

MONDAY, August 16, 1824, at the CIRCUS,

Where it will continue until SATURDAY the 21st,

Admission to the Exhibition at the CIRCUS, Ladies and Gentlemen, 1s.—Servants and Children, 6d.; open each Day from Nine o'Clock in the Morning until Eight at Night.

On MONDAY, AUG. 23,

At FOUR o'Clock in the Afternoon, Mr. GREEN purposes making his

20TH AERIAL VOYAGE,

FROM THE

Yard of the Gas-Works,

SHREWSBURY.

half-filled by 'the assemblage of fashionable and highly respectable personages of both sexes', many thousands having taken to vantage points round about, as far afield as Clive, Grinshill and Loton. The balloon rose slowly from the pit and was held momentarily in a state of buoyant equilibrium while Green checked his apparatus, which included 'two beautiful flags', a map of the county, a basket containing a pigeon ('for scientific experiments') and a load of junk, the personal possessions of friends and well-wishers, carried aloft 'with view of furnishing to their proprietors memorials &c. of Mr. G's trackless journey'. The car, a slender wicker-work boat-shaped gondola 2ft wide, 5ft long, 2ft 3ins deep, embellished with signs of the Zodiac, elements and seasons, was attached. At 4.27 pm the small pilot balloon was released. Three minutes later, 'waving his flags amid the cheers of the spectators, and the Bands of Music playing God Save the King, Mr. Green took his adventurous flight in the most grand and satisfactory manner possible'. There was a quick squint at the weather-vanes of churches to ascertain wind direction; the pigeon was released at 4,000ft (but it never returned to its Shrewsbury loft). Rising still higher, magnificent views of the Welsh hills were obtained as far as Cader Idris. The balloon was taken west of the Wrekin 'on which he observed several groups of persons; he put out and waved his flag, on which he distinctly heard the cheers of the parties' perhaps a bit of journalistic licence, as by his barometer Green was an estimated 2–2$\frac{1}{2}$ miles high! At 6.50 pm the Royal Coronation Balloon landed at Skimblescott, in Monkhopton parish, Corve Dale, being pulled gently down and made fast by farm labourers in the fields below. At the insistence of R. H. G. More, Esq., he was made welcome at Larden Hall across the river until a post-chaise and four, ordered earlier with some remarkable prescience from Much Wenlock, arrived to convey astronaut and balloon thence, arriving shortly before midnight to an ecstatic welcome at the 'White Hart'.

Green returned to Shrewsbury the following morning. Two days later an indignant correspondent to *Eddowes's Journal* suggested that all those who, for one reason or the other, did not purchase tickets for the gas-yard grandstand and 'to whom the money can be no object or a very trifling one', should make donations at the printers towards the further remuneration of Mr. Green 'for the risk, anxiety and expense, which have been necessarily incurred by sending this floating wonder high through the fields of air. . .'

Following the success of the occasion it was announced that Mr. Green, 'on the suggestion of many Gentlemen of the town and vicinity… purposes making a second ascent from Shrewsbury during Race Week if he can make the requisite arrangements or [find] a suitable point of ascension'. Charles Green duly revisited the town on Friday, 24 September. In a more streamlined operation his balloon was inflated in fifty minutes, a severe strain on a gas-works that as yet only produced 30,000cu. ft a day! The balloon was then 'walked' over the wall into the garden of one Mr. Kirkham for viewing and was launched at 3. 10 pm The heat of the sun's rays reflecting off the lower layers of cloud caused some discomfort, so that much of the trip was spent at an altitude of 3–3$\frac{1}{2}$ miles, often in cloud so dense he could not see across his tiny car. There was a strong SW wind blowing and he made a rather bumpy landing at 5pm, four miles beyond Knighton and thirty-four miles from Shrewsbury. On the ground the 'Committee of Gentlemen' had got their act together and spectators were forewarned that 'friends with tin boxes… distinguishable by a white silk ribbon with a blue device of the balloon worn in the lapel of the coat' would pass amongst them. To lessen the blow and overcome criticism 'half of the sum collected will generously be given to the Royal Shropshire Infirmary'.

To Charles Green these were 'apprentice flights'. He would go on to become the greatest British aeronaut and one of the most skillful and successful of balloon pilots. Cool in emergencies, with lightning-quick reactions, he was a martinet in his balloon, imposing an iron discipline upon his passengers regardless of rank or ticket price. This was seldom resented, as it inspired a confidence seldom misplaced. By 1835 and one balloon later, he had made over 200 ascents, logging 240 flying hours, and covering some 6,200 miles. By 1844 he had made 299 flights, thirteen at night, and had carried 548 passengers, twenty-eight of them women. Despite advancing age (he was born in 1785) Green continued to make regular ascents, obviously aiming to achieve a personal target of 500 flights. His 'positively last appearance', advertised as the 500th ascent, was scheduled for 12 September 1852, although opinions vary as to whether it was actually something between his 504th or 526th! In 1866 Green became a founder member of the (Royal) Aeronautical Society and died suddenly of heart failure in March 1870 at the ripe old age of 85!

By contrast the two most accident-prone aeronauts to visit Shrewsbury, Welshpool and other border towns were the Grahams, husband and wife. They had no need to indulge in spectacular supporting stunts to draw large crowds. Their very appearance guaranteed sensation and/or debacle — ripped or leaking balloons, unceremonious dunkings in river, sea or gravel pits, capsizing, dislodged cornices, copings and chimney pots etc. They were masters of the art of roof-skimming with dangling grapnel! Unlucky in the practicalities of ballooning, they were both fortunate enough to survive all accidents and die in their beds. George Graham, being the elder of the pair, gave up ascents in 1838, but the amazing Margaret, interludes of child-bearing permitting, would soldier on for another thirty years!

George Graham descended unheralded on the unsuspecting town of Shrewsbury on the night of Tuesday, 18 September 1838. He declared his intention of making his 248th ascent the following Thursday from the vastness of the Racecourse, 'preferring to gratify the public by an open ascent'. This decision drastically reduced the takings despite an announcement that Margaret Graham would make an ascent on the first day of Race Week, which roused interest and helped swell the crowds on Monkmoor fields. Monies taken at the Gas Office during the course of inflation did not cover the cost of the gas. On the other hand, as the *Salopian Journal* rather unkindly observed, '… the inhabitants of the county would have been deprived of the most entertaining spectacle they have witnessed for years!'

After waiting for strong winds to subside, the balloon was guided, an hour late, from the gasworks to the river bank, striking roofs of surrounding buildings and threatening to tip the aeronaut from the car. This was the easy bit! As the local news-sheet records: '. . . the difficulty of piloting the balloon across the Severn, and carrying over the men who held the ropes, whilst the vast machine, high in the air, was struggling against them to get free, appeared so perilous, and threatened so many difficulties, that the vast multitude who filled every spot of the neighbourhood, were held in breathless expectation till the first boatload of men, with the leading rope, were safely landed on the opposite side, and the balloon slowly wended onwards to the Race Course, where it was safely brought to ground in front of the Grand Stand'.

Ascent was delayed for a short while while a slightly intoxicated 'volunteer assistant, who leaped into the car and insisted on sharing the perils of the exploit', was firmly removed. Pantomime over, Mr. Graham made 'a beautiful and majestic ascent' at 4.55 pm The air currents carried him out over the Severn towards Uffington, Rodington, and Longdon-on-Tern, the balloon finally landing thirty minutes later in a meadow at Long Lane, Eyton-upon-the-Weald Moors, about two miles from Wellington. Plenty of assistance was to hand to bring the balloon safely to earth. A post-chaise being speedily forthcoming, the intrepid aeronaut was back in Shrewsbury before 9 pm The next day notices were posted soliciting subscriptions to cover some of his costs. A group of gentlemen from Welshpool, having witnessed the spectacle took it upon themselves to invite the Grahams to make an ascent from that town ten days later. Possibly in an attempt to steal some of Shrewsbury's thunder, Mrs. Graham would be

GRAND TREAT.

MONDAY, OCT. 2nd, 1848,
BY THE CELEBRATED ÆRONAUT
MR. WADMAN,
WHO WILL SUBMIT FOR INSPECTION THE CURIOUS MONSTER ÆRIAL MACHINE,
AND WILL MAKE HIS 66th
GRAND ASCENT
AT FOUR O'CLOCK IN THE EVENING OF THE SAME DAY, IN HIS BALLOON
THE "RAINBOW,"
Some idea of the MAGNITUDE of this STUPENDOUS MACHINE may be formed by the fact of its taking upwards of
700 SQUARE YARDS OF SILK
To construct it, and the net-work of the same contains more than 40,000 FEET of fine flexible CORD

The Balloon will be inflated by the CHESTER GAS COMPANY, Manager Mr. T. PROUD, who will lay on large Pipes for the purpose.
For the gratification of Parties desirous to WITNESS THE PROCESS OF INFLATION, and of closely inspecting the ÆRIAL MONSTER, the CURIOUSLY CONSTRUCTED CAR, and its various appendages, the DOORS WILL BE OPENED AT TWELVE O'CLOCK, and all arrangements connected with the Ascent, will be under the superintendance of

MR. R. GREEN

Poster advertizing the flight of R. Wadman, the Bristol aeronaut, from Chester in October 1848. [Chester Libraries]

summoned from London and was tentatively pencilled in to make her first ascent from the Montgomeryshire town.

On Friday, 28 September 1838, over 1,000 people were gathered in the screened-off enclosure at the Bowling Green, Welshpool, whence the Grahams' balloon had been towed the quarter of a mile across town from the gasworks. The small town's gas supply was over-taxed. Inflation had commenced at 10 a.m.the previous day and was barely completed thirty-two hours later. Shortly after 5 pm Mrs. Graham took her place in the car and had lift off, but obviously buoyancy problems were anticipated as grappling iron, ballast, seats and other non-essentials had been hastily removed from the basket. Cheers quickly turned to cat-calls, and there were ugly scenes as the balloon rose only 80ft from the ground before coming to earth again. It was announced that 'the balloon was incapable of ascending with Mrs. Graham's weight; a cry was raised that the balloon ought to ascend, which was eventually acceded to, and off it went, without any one in the car, in fine style, soon attaining an altitude of 1,000 to 2,000ft'. A very suspicious journalist revealed 'as an undisputed fact that papers were attached to the balloon previous to Mrs. Graham getting into the car, with directions where the balloon was to be sent to when found'. He adds darkly, 'We refrain from comment; the public must judge what such previous arrangements could be intended for!'

Paying spectators had to be compensated for their disappointment and the organising committee insisted on Mr. Graham fulfiling his contract by making a gratuitous ascent the following week. Failing this a sum not less than 'the amount he was engaged for will be given to some Public Charities'. Meanwhile a phaeton from 'The Bear' inn was

Mrs. Graham's

ASCENT.

THE Nobility, Gentry, and Inhabitants of Shrewsbury and adjoining Neighbourhood, are most respectfully informed, that

MRS. GRAHAM

The only English Female Aeronaut,

WILL MAKE HER

First Ascent,

FROM ADJOINING THE GAS-WORKS, SHREWSBURY,

ON SATURDAY NEXT, OCTOBER 6th, 1838,

PRECISELY AT THREE O'CLOCK IN THE AFTERNOON.

Admission to view the Inflation at the Gas-Works, with return Tickets to the place of Ascent, 2s; Children, half price. To the Ascent only 1s.

After the Welshpool debacle in the September, Margaret Graham finally made her first balloon flight from Shrewsbury on 6 October 1838. [SRCC].

hired to follow the balloon, which finally came to earth on the 1,000ft ridge at Lawnt-y-gwynt, Llanerfyl, at about 6.30pm, when a labourer in the harvest field, 'supposing the aeronauts were inside the massive machine, with a reaping hook cut a hole in it and cried out, requesting them to come down and get out'. The balloon was recovered and conveyed to Shrewsbury, where repairs 'including 23 yards of silk, varnish, preparation, tailors etc.' amounted to £11 7s..

It is pleasing to relate that after such a debacle, Margaret Graham's ascent at Shrewsbury on Saturday, 6 October, launching her career as 'Her Majesty's Aeronaute' or 'The Only English female Aeronaut', went without a hitch. She carried as a passenger one Thomas Bertenshaw of that town, 'an anxious aspirant for aeronautic fame'. The prevailing calm made for a slow ascent and progress. They drifted ten miles in fifty minutes, first in a north-westerly direction over Berwick House and The Isle, then circling southwards where they thought it prudent to descend to avoid the numerous stubble fires. They made a perfect landing at Halston Farm, Pontesford, where they were hospitably entertained until Mr. Graham arrived with a chaise. They were back in Shrewsbury by 8 pm The *Salopian Journal* concluded, rather lamely for that paper, '… it is to be hoped that the proprietor of the balloon will receive the remuneration which his repeated endeavours to gratify the public curiosity so well deserve'.

Just add an engine

Improving on rather shaky beginnings, the latter half of the nineteenth century turned out to be the heyday of ballooning, due in no small measure to the pioneering efforts of Edward Spencer and Charles Green. They founded a balloon manufacturing business in London producing balloons not only for themselves as the country's leading aeronauts but also for the growing number of 'professional' balloonists who eked out a precarious living on the entertainment circuit. To these one could add the well-heeled young gentlemen who, in the natural progression of things, took on board ballooning as the 'in' sport or pastime before abandoning it in favour of the motor-car and then the aeroplane.

Spencer's son, Charles Green Spencer (so christened after his god-father) also went into the ballooning business. His company, C. G. Spencer & Sons, of Highbury, established a virtual monopoly in the manufacture and display of both gas and hot air balloons. From 1907 Short Brothers, perhaps better known as aeroplane builders, beginning with the Short-Wright bi-plane of 1909, would provide stiffening competition. But for the moment, from 1894, Charles Spencer and his sons, Stanley, Arthur and Percival would dominate ballooning events in the provinces, certainly in Shropshire and 'The World's Wonder Show' (otherwise 'Shrewsbury Floral and Musical Fete). As early as 1877 the latter's organisers had realised that it 'could only survive by offering better entertainment than anyone else'. They had almost exhausted the gamut of trapeze artists, tight-rope walkers (with or without bear and/or cooking an omelette), 'equilibriumists', contortionists, the 'White Helmets' (aka the Royal Corps of Signals display team), knife throwers, trick cyclists, the 'Wall of Fire' and show jumping etc., etc.

In 1880 another novelty was tried — 'Grand Balloon Ascents' by 'Captain' James A. Whelan of Huddersfield, an aeronaut of some seven years experience, averaging twenty-two ascents a year. This certainly drew the crowds and the Show's organisers would persist with gas balloons and even primitive airships until the outbreak of war in 1914. Whelan flew a succession of named balloons — *Excelsior* (25,000 cu. ft.), *Duke of Edinburgh*, [Queen] *Victoria*, and *Shrewsbury* (Mk.II at 60,000 cu. ft.) — usually carrying one or two passengers (three to six in *Shrewsbury*'s basket). The first year of his contract he flew solo,

Tethered flights from the Quarry, Shrewsbury c1909, possibly Condor *(left) and* Invincible. *[SRRC]*

establishing his credentials as it were, before persuading Mr. Belton, manager of Shrewsbury Gas Works, to take a 20-minute flight, albeit in stormy conditions, on the last day of the 1881 Show. This was a tacit acknowledgement of the latter's co-operation and efficiency in providing pumps and a temporary extension of the town's gas main from Town Walls into the launching enclosure in The Quarry. This broke down any reservations that the public might have had regarding balloons.

Whelan was also an astute PR man. His last flight from the Quarry on Thursday, 16 August 1883, saw the *Duke of Edinburgh* carried out over The Column and away towards Wellington. He tacked about The Wrekin for some considerable time before passing at 5,000 ft. over Ironbridge to the cheers of colliers. The prevailing wind took the balloon over Shifnal and the suburbs of Wolverhampton were in sight before a landing was effected in a turnip field at Bishton on the earl of Dartmouth's estate. The balloon and basket were carried by eager farm labourers into an adjacent pasture field and cart and horses quickly found to carry the balloon and tackle to Albrighton Station. Pilot and a much relieved passenger arrived back in Shrewsbury 23.23hrs, but not before expressing their thanks by giving two captive ascents to the farm-workers' children!

Whelan's intendancy at Shrewsbury was abruptly terminated under tragic circumstances in 1893. It was his thirteenth appearance at the Show and his 315th balloon ascent. On Wednesday, 23 August, the *Victoria* ascended with one passenger from The Quarry amphitheatre and was immediately whisked away eastwards towards The Wrekin. The landing in open country at Cruddington, near Wellington was a bumpy one, the grappling hook twice failing to hold. The basket was dragged and rolled along the ground, Whelan sustaining a broken pelvis as he was thrown violently against the side of the car. This was only his second accident in some twenty years of ballooning, the first being in 1881 when, after his Shrewsbury engagement, he went on to a 'Sports Day' at a small village near Stafford.

A windy day and a tiny, cramped enclosure caused the balloon to 'luff' about the heads of pardonably apprehensive spectators who fended off the crazy sphere with hands and sticks, unwittingly holing the fabric in the process. Misshapen and unstable, the balloon literally rolled out of the ground and into the next field becoming entangled in advertising hoardings, telegraph wires and sundry chimney stacks. The intrepid aeronaut spent three weeks confined to bed. But a broken pelvis was a different matter. A rough ride in a farm cart to the Royal Salop Infirmary did not help. Complications set in and the 53-year old showman died on 2 September. Yet his show went on. His assistant called upon the services of a Mr. Jackson of Derby and on the Thursday the latter's smaller balloon, the *Princess Mary*, took off from the showground dead on schedule, but landing, perhaps prudently, 'only a few miles from Shrewsbury'.

The gap in the 1894 programme was filled by Stanley and Percival Spencer, 'aeronauts to Crystal Palace', who had been drawing the crowds by 'exciting ascents' at Cardiff. For the 1895 Show balloons were supplied and flown by Mr. Beatson of Huddersfield (200 ascents to his credit) and Mrs. Beatson (8 ascents). Obviously a lady pilot cut little ice with the Show's organisers — or was it the fact that the balloon flights (to Hadnall and Stanton upon Hine Heath) were not all that venturesome? Whatever, the 1896 ballooning contract or sponsorship was awarded once more to the Spencers who would retain it until 1914. They celebrated by displaying their latest novelty, 'a Flying Fish' balloon, perhaps heralding the company's 'diversification' into early airship design. One is mindful that balloons shaped to resemble and promote commercial products are not confined to present day hot-air balloons — John Gosnell's 70ft French-built 'Cherry Blossom' perfume bottle was floating ('tottering' might be a better word!) about Pennine skies as early as May 1893!

Basically the Spencers maintained the programme very much as

before, with two balloons 'in steam', giving captive flights up to 500ft whatever the weather for 5s., taking people 'on a voyage of the Fete' as it were. These did not always go to plan. For example, at 1730hrs on Wednesday, 22 August 1906, *Wulfruna* which had been giving captive flights for most of the afternoon, suddenly gave a violent upward jerk and snapped its tether just below its basket. Five delighted passengers, including the vicar of Bridgnorth, embarked on a once-in-a-lifetime voyage at 10,000ft out over Newport, to eventually make a perfect landing at Knightley Eaves, near Eccleshall, Staffs.

Barring such mishaps, at 16.30 hrs. and 19.00 hrs. the balloons slipped their cables to make scheduled free flights — but to unscheduled destinations. At 20.00 hrs. came the grand finale — the ascent of dozens of small 'fire balloons' with fireworks and magnesium lights, highly dangerous but no doubt effective. Under new management the balloons increased in size, passenger carrying capacity doubled, and men of substance vied with each other to be carried aloft and afar, especially after 1909 when the traditional staggered last flights of the day gave way to dramatic long distance cross-country 'balloon races'. Only four females are recorded as venturing on these flights. The first was a young lady from Huddersfield who 'was warmly applauded for her courage'. She needed it! On 23 August 1887 she was the sole passenger; the flight was short — only to Whitchurch via Wem — but it was a rough journey and the landing even bumpier, achieved only on the second attempt. Other ladies flew in 1902, 1904 (in the Spencer airship!) and 1905.

While Shrewsbury's balloons were purely entertainment, almost fairground attractions, with large crowds watching the inflation process and launch as much in the hope of witnessing some mishap as for the thrill of the flight itself, occasionally there was some scientific or academic input into these flights. On Thursday, 18 August 1898 a captive Wulfruna was put at the disposal of the Rev. John Bacon, a well-known amateur balloonist and writer on aviation matters — and for once without his daughter, Gertrude — to carry out 'atmospheric research' along with scientists from Cambridge University. These, however, pale into insignificance with the world altitude record balloon ascents made by James Glaisher and Henry Coxwell at Wolverhampton in 1862. More frustratingly one learns of the activities of Mr. H. Cove who for many years either side of 1900 covered the Flower Show for the *Gardeners' Chronicle*. Apparently he was a frequent flier and always took his camera aloft, securing the first aerial photographs of the town and environs. One wonders what happened to this important archive or if some prints were actually published? John Bacon was also in the habit of taking aerial photographs, but not, as far

Prototype Spencer airship over Crystal Palace, 1902.

The larger Spencer airship which courted disaster at Shrewsbury Flower Show, 1906. [SRRC]

as can be ascertained, on the occasion of his Shrewsbury ascent.

In 1899 J. H. Parsons, 'gas manager', Oswestry, was an interested passenger in the Shrewsbury but whether a voluntary one is open to debate as the balloon's tether was slipped as a precautionary measure as the wind freshened and, whether they liked it or not, passengers and pilot were whisked off over Baschurch, Petton and Ellesmere at over 30mph. The later presence of the GWR goods agent from Oswestry would seem to suggest that both these men were in Shrewsbury in their professional capacity, assessing the logistics of ballooning possibly with view to adopting it as an attraction in their own town.

In 1904 novelty momentum would be maintained by the appearance of Stanley Spencer's 'airship' described in the promotional blurb as 'most likely to be the only one flying in Britain at the time'. This flew only when sponsorship was forthcoming, but on the first day of the Shrewsbury Show it was not even unpacked so stormy was the weather! But although the Spencer 'airship' signified a rather wobbly intermediate stage in man's progress from an aerostatic globe floating free at the mercy of prevailing winds to the aeroplane, a machine which could take off under its own power and (hopefully) proceed to a destination as directed by a pilot, the balloon did not yield aerial supremacy without a struggle.

On 15 February 1910 the Irish aeronaut, John Dunville, in the *Banshee*, repaired after its envelope had been badly torn a month earlier, crossed the Irish Sea west—east from Dublin to Macclesfield. It was still very much a novelty to the gawping crowds at South Stack lighthouse, Llandudno, Colwyn Bay, Rhyl and Mostyn as it crossed the tip of the Clwydian Range at 2,000ft before heading out over Neston, Parkgate and Frodsham, If he had landed on the holiday coast, crossing achieved, willing and knowing hands would have walked him down. Balloons had long been a feature of places of entertainment in Congleton, Oldham, Salford, Stockport, Stalybridge, Manchester (the old Racecourse and Bell Vue) and Rochdale Infirmary Galas. In more rural areas the landing of a balloon was still the experience of a lifetime, a moment when callow farm lads became men! Thus on Wednesday, 20 August 1884, as a finale to the first day of that year's Flower Show, the balloon *Shrewsbury* — as yet a mere 35,000 cu. ft.'but neither shapely nor graceful' for all that' — took off at 17.55 hrs. from The Quarry, much later than scheduled owing to inflation problems. It was of necessity a short flight, following the railway line

to Upton Magna and Walcot before unceremoniously dumping its passengers in a field at Wheathill Farm. As a tired old hack, writing for the *Shrewsbury Journal* commented — one can almost imagine a slight sneer and curl of lip — 'the natives who ran to the spot were considerably astonished and a farm labourer who was asked to take hold of a rope gazed in a state of trembling bewilderment at the balloon, evidently thinking it was a live animal wonderfully and fearfully made'.

Dunville's flight could have carried on much further, but the inhospitable Pennines loomed. Discretion got the better of valour and an orderly descent was made at Macclesfield. The moors had long entered balloonists' mythology. On 30 September 1859 a 'celebratory' balloon from Stockport, with Lt. Peter Fosbrooke of the Cheshire Yeomanry as passenger, came down on Greenfield Moor, north-east of Mossley. It took nearly four hours of wandering in the gathering gloom before the aeronauts secured help from an isolated farmstead. Little daunted by his first flight, Lt. Fossbrooke secured a place on other balloon flights in the years following. Did he cherish a secret ambition to become his regiment's first Balloon Officer?

The Spencers also operated named balloons out of Shrewsbury— *Carnation, Victoria, Wulfruna, Shrewsbury* (several models, with *Mk.III* emblazoned with the borough's coat of arms and motto), *City of London, City of York, Graphic, Aeroclub N° 1* (appearing in 1903, the year in which the Aero Club introduced Saturday races from the Ranelagh polo ground at Barnes), *Norfolk, Condor* (with its coating of aluminium 'dope' or paint), *Invincible* and *Enchantress*. The largest was *Shrewsbury Mk.V*, specially constructed for the 1911 season. At 75,000 cu. ft. it had a padded car to lessen chances of injury in bumpy landings or turbulence, and upholstered seats 'for long distance work'. It could carry eight passengers.

Generally navigation was not much of a problem. Bearings could be taken from the railway lines radiating out from Shrewsbury, Wellington, Oswestry and Whitchurch, and landmarks such as the Wrekin, the Breiddens and the Dee and Severn. Strangely pilots did not seem to bother carrying Ordnance Survey maps, possibly because the individual sheets of the then current 'Second' and 'Third Editions' of the One-inch OS map covered such a small area (12 x 18 miles). Some of the long distance flights made would have needed over a dozen maps, to say nothing of the problem of consulting them in strong winds

BALLOON FLIGHTS FROM SHREWSBURY, 1880–1913
↘ = landed

1880
18 August: J. A. Whelan, solo, in *Excelsior* (25,000 cu. ft), 6 miles, 12,000 ft., 30mins, ↘ Ford Heath.
19 August: J. A. Whelan and assistant, in *Duke of Edinburgh* (30,000 cu. ft).

1881
17 August: J. A. Whelan in *Duke of Edinburgh*.
18 August:J. A. Whelan and Mr. Belton (Shrewsbury Gas Works) in *Duke of Edinburgh*, 20mins, Column–Hanwood ~ Sutton.

1882
16 August: J. A. Whelan in *Excelsior*, 4,000 ft., 35mins, Chilton-Cound ↘ Eyton on Severn.
17 August: J. A. Whelan, one passenger, in *Duke of Edinburgh*, 45mins, Sundorne–Haughmond Shawbury.

1883
15 August: J. A. Whelan (128th ascent) in smallest balloon (18,000 cu. ft), 5,300 ft., 24mins, Wrekin ↘ Church Farm, Buildwas.
16 August: J. A. Whelan, one passenger, in *Duke of Edinburgh*, 4,000 ft., Column–Wrekin Ironbridge–Shifnal– Pattingham ↘ Bishton.

1884
20 August: J. A. Whelan, three passengers, in *Shrewsbury* (35,000 cu. ft.), largest seen in the area, Upton Magna–Walcot ↘ Wheathill.
21 August: J. A. Whelan, three passengers, becalmed 12 mins., 2,500 ft., along S&C railway ↘ Oldheath.

1885
19 August: J. A. Whelan, two passengers, ↘ Dorrington. Mr. Beatson, Huddersfield, veteran of 200 flights, supervised captive flights.
20 August: J. A. Whelan, one passenger, strong wind, out to Radnorshire border, Stannage Park–Bucknall ↘ Craven Arms.

1886
18 August: J. A. Whelan, six passengers, in *Shrewsbury* (60,000 cu. ft.), 3,000 ft., 2¹/2 hrs., Uffington–Atcham–Wrekin–Wellington–Haughton ↘ Shifnal.
19 August: J. A. Whelan in *Shrewsbury*, Condover–Pitchford–Titterstone Clee Hills ↘ Squirrel Lane, Snitton.

1887
22 August: J. A. Whelan, two passengers, in *Victoria* (30,000 cu. ft), 45 mins., Wrekin Wellington–Coalbrookdale ↘ Haughton Hall, Staffs.
23 August: J. A. Whelan, lady passenger, along railway to Wem ↘ Whitchurch (2nd attempt).

1888
22 August: J. A. Whelan (236th ascent), two passengers, in *Victoria* (17th ascent), 1¹/4 hrs. ↘ Orton Hall, Staffs.
23 August: J. A. Whelan, two passengers, in *Victoria*, 83 mins., Wem–Whitchurch–Mill Bank ↘ Fenns Gorse (Maelor).

1889
21 August: J. A. Whelan, one passenger, in *Victoria*, 50 mins. ↘ High Ercall.
22 August: J. A. Whelan, one passenger, in *Victoria*.

1890
20 August:J. A. Whelan, one passenger. *Shrewsbury* (53,000 cu. ft.) prepared, but bag ripped during inflation process. Smaller balloon readied in one hour, 25 mins., 5 miles ↘ Uckington.
21 August: J. A. Whelan, one passenger, in *Queen Victoria* [sic] ↘ Copnall (2 miles from Stafford). Smaller balloon also launched.

1891
19 August: No details available.
20 August: No details available.

1892
17 August: J. A. Whelan, one passenger, in *Victoria* ↘ Meole Brace.
18 August: J. A. Whelan, two passengers↘ Harlescott.

1893
23 August: J. A. Whelan, one passenger, in *Victoria*, Wrekin ↘ Crudgington. Bumpy landing, Whelan fractures pelvis, dies later from complications.
24 August: Mr. Jackson, Derby (Whelan's assistant) in *Princess Mary* ↘ 'a few miles from Shrewsbury '.

1894
22 August: Percival Spencer, four passengers, in specially made *Carnation* (32,000 cu. ft.), Hadnall-Wellington ↘ Ercall Park.
23 August: Percival Spencer, solo (bad weather), Dinthill↘ Ford.

1895
20 August: Mr. Beatson, Huddersfield, three passengers, 4,000 ft. ↘ Hadnall.
21 August:Mr. Beatson, two passengers, 5,400 ft., 45 mins., Sundorne-Moreton Corbet ↘ Stanton upon Hine Heath.

1896
19 August: Mr. Williams (assistant to C. G. Spencer), one passenger in 'Flying Fish Balloon', 10,000 ft. ↘ Abdon, Bridgnorth.

20 August: Percival Spencer, one passenger, in *Carnation* ↘ Berrington.

1897

18 August: C. G. Spencer, one passenger, in *Victoria*, poor weather ↘ Bayston Hill.

19 August: Arthur Spencer, two passengers, beyond Wrekin at late hour, no further details.

1898

17 August: 4 pm Arthur Spencer, two passengers, in *Victoria*, 3,000ft ↘ Ford. 7 pm Arthur Spencer, two passengers in *Wulfruna*, 4,000 ft., Shrawardine-Alderton ↘ beyond.

18 August: 4 pm Arthur Spencer, two passengers, in *Victoria* ↘ between Cruckton and Hanwood. 7pm Percival Spencer, two passengers, in Shrewsbury, Wollaston ↘ near Breiddins, walked over Welsh border!

1899

23 August: 4pmArthur Spencer, two passengers, in *Victoria*, 2,500 ft., 6 miles, crossed Severn five times ↘ Mytton. 7pm Percival Spencer, one passenger, in *Shrewsbury*, crossed Severn five times ↘ Fitz.

24 August: 4pmArthur Spencer, one passenger, in *Victoria*, 35 miles ↘ in a strawberry field at Farndon. 7pm Percival Spencer, three passengers (including J. H. Parsons, Oswestry Gas Works) in *Shrewsbury*, 30 m.p.h. winds, Leaton–Baschurch–Petton–Ellesmere ↘ in a meadow, Bangor-on-Dee.

1900

22 August: Balloons: *Victoria* (40,000 cu. ft), *City of London* (24,000 cu. ft), *Shrewsbury* (new, 35,000 cu. ft). 4pm Percival Spencer, three passengers (including H. Cove of the *Gardeners' Chronicle*, a regular flier from the quarry, who took some of the earliest aerial shots of Shropshire), in *Victoria*, Wellington–Wrekin ↘ Flashbrook Manor, Staffs. (tangled with barbed-wire fence). 7pm Arthur Spencer, three passengers, in *Shrewsbury*, heavy rain ↘ Yorton.

23 August: 4pmArthur Spencer, one passenger, in *City of London*, 5,000 ft., Oswestry Ellesmere in one hour ↘ Maes-y-llan, Erbistock. 7pm Percival Spencer, two passengers, in *Shrewsbury*, ↘ Cross Lanes, St. Martins.

1901

21 August: 4.30pm Capt. Smallbone, one passenger (Mr. Littleboy, Pen-y-garth, Wrexham), in *Shrewsbury* (25,000 cu. ft), 2,000 ft., north-west ↘ Montford. 7pm Percival Spencer, two passengers, in *City of York* (40,000 cu. ft), north-west along Severn ↘ Endon House, Montford Bridge.

22 August: 4pm Arthur Spencer, one passenger (Mr. Bernard Head, Ellesmere) in *Shrewsbury* ↘ Lyth Hill (decided not to gate-crash Lythwood Hall garden party!). 7pm Percival Spencer, four passengers, in *City of York*, Bomere Pool–Condover Church Stretton 'rough country' ↘ Stretton.

1902

20 August: 4pm Arthur Spencer, three passengers, in 'splendid new *Shrewsbury* (40,000 cu. ft) carrying borough coat-of-arms and motto, Wellington–Dawley mining districts Shifnal. 7pmPercival Spencer in *Graphic*, Wrekin–the Ercall ↘ Lawley Bank.

21 August: 4pmArthur Spencer, three passengers (including a lady), in *Shrewsbury*, 1 hour, ↘ Wellington. 7pmPercival Spencer, four passengers (including A. C. Minshall, The Cross, Oswestry jeweller), in *Graphic*, 4,100 ft., along Severn ↘ one mile from Wellington.

1903

19 August: 4.30pm Arthur Spencer, two passengers, in *Shrewsbury* (33,000 cu. ft), 5,000 ft. in cloud, Wellington–Newport ↘ 2 miles SW of Stafford. 7pm Percival Spencer in *Aeroclub N°. 1* (45,000 cu. ft.), Newport in 40 mins. ↘ Chetwynd Grange.

20 August:7pm only. Steady rain, high winds. Percival Spencer, rough ride ↘ Broome, near Aston on Clun.

1904

17 August: Balloons: *Shrewsbury* (40,000 cu. ft.), similar back-up, and an airship. 7pm only. Unfavourable weather. Stanley Spencer, three passengers, in *Shrewsbury* ↘ Isombridge, near Walcot.

18 August: 7 pm, bad weather abated. Percival Spencer, two passengers, (probably in *Shrewsbury*), Wellington—along LNWR line to Donnington ↘ Lubtree Park. 7.30pm Stanley Spencer, in airship, with Gertrude Bacon, daughter of Rev. J. H. Bacon and writer on aviation. Risky ascent, touching tents and trees; upper currents too strong for engine ↘ Admaston.

1905

23 August: Balloons: *Shrewsbury* (45,000 cu. ft.), *Norfolk* (35,000 cu. ft.), *Carnation* (35,000 cu. ft.). 4pm Percival Spencer, E. R. Cooke (balloonist), two passengers, in Shrewsbury, one of longest flights, 3,500 ft., Wellington– Wrekin–Shifnal—raced (40 m.p.h.) GWR express into Wolverhampton–Walsall–Birmingham ↘ Doe Bank Farm, Great Barr. 7pm Arthur Spencer, two passengers, Wrekin ↘ Huntington, Little Wenlock.

24 August: 4pmE. R. Cooke, three passengers, in *Carnation*, 4,000 ft., 1¼ hrs. ↘ Acton Reynald. 6.45pm Percival Spencer, three passengers (including a lady) ↘ Albrighton, near Shrewsbury.

1906

22 August: 4pm Percival Spencer, two passengers, in *Shrewsbury*, 1½ hrs., Haughmond Hill Peplow ↘ Market Drayton (?Stoke Grange). 5.30pm *Wulfruna* (45,000 cu. ft) broke loose on captive flight, with Arthur Spencer and five delighted passengers, 10,000 ft., Newport ↘ Knightley Eaves near Eccleshall, Staffs.

23 August: 4pmArthur Spencer, two passengers, hazy, kept low altitude ↘ Llanymynech. 7pm Percival Spencer in *Wulfruna*, 'on cross-country steeplechase'—along disused 'Potts Railway'–Breiddins–Tanat valley ↘ Llanyblodwel.

1907

21 August: 4pm Arthur Spencer, four passengers, Atcham–Cressage–Ironbridge–Worfield Gatacre–Enville Racecourse ↘ Stourton Fields between Dudley and Stourbridge. 6.30pm Percival Spencer, two passengers (including Mr. A. Poly-Didier,

Gresford), 3,600 ft., Much Wenlock–Morville ↘ Bridgnorth.

22 August: No details available.

1908

No details available.

1909

18 August 4pm Percival Spencer, two passengers, in *Condor* (45,000 cu. ft., coated aluminium paint), 6,000–7,500 ft., 40-50 m.p.h. winds, Bridgnorth–Redditch–Stratford-upon-Avon ↘ Tysoe (west of Banbury). 6.30pm Sydney Spencer, three passengers (including T. Hogben, Chester), in *Invincible* (35,000 cu. ft.), 30 mins. ↘ 'in a snug little valley the other side of the Wrekin'.

19 August: 4pmSydney Spencer, three passengers (including T. Hogben, Chester) in *Shrewsbury* (45,000 cu. ft.), determined to beat long-distance record ↘ Newark. 6.30pm Percival Spencer, six passengers (large Oswestry contingent — G. H. Aston, GWR Wharf; E. L. Morton, Morda; A. C. Minshall, silversmith), in *Condor,* Wellington–Newport—no news of descent.

1910

17 August: 4.30pmHerbert Spencer, four passengers, in *Shrewsbury*, 35 miles, 3,900 ft., Wellington–Newport ↘ near Stafford. 6.30pmSydney Spencer, two passengers, in *Carnation*, 70 mins., 3,200 ft. Wellington–Penkridge ↘ Rugeley 7.40 pm

18 August: First of 'balloon races' *i.e.* simultaneous releases. No details known.

1911

23 August 4.30pmHenry Spencer, eight passengers, in specially constructed *Shrewsbury* (75,000 cu. ft.) with upholstered car and seats for long-distance work, 2¹/4 hrs. becalmed, then Leaton, Berwick (picked up angler who knew the country) ↘ Marton Hall, Baschurch. 6.30 pm Percival Spencer, one passenger, in *Invincible* (35,000 cu. ft.), 3,500 ft. ↘ Newton Farm, Berwick.

24 August: 5.30-5.34pm 'Race': Percival Spencer and 'Mayor's party' (8) in *Shrewsbury*, Henry Spencer, four passengers in *Invincible*. Both High Ercall–Crudgington Newport–Peplow. *Shrewsbury*, 30 miles, 25 m.p.h. ↘ Maer, Whitmore. *Invincible* ↘ 20 mins. and 3 miles further.

1912

21 August: 4.30pm Harry Trueman, three passengers in *Shrewsbury* (35,000 cu. ft.), 6,200 ft., 2 hours, Wellington–Newport-Stafford ↘ Colwich. 6.30pm Henry Spencer, two passengers, in *Enchantress*, 1¹/2 hrs. ↘ Fernhill, Staffs. (3 miles north of Newport).

22 August: 5.30pm 'Race'. No further details available.

1913

20 August: 4.30pm Harry Spencer, two passengers, in *Invincible* (35,000 cu. ft.), 5,300 ft., Buildwas ↘ Wyke (west of Broseley). 6.30pm Henry Spencer, one passenger, in *Shrewsbury*, 7,000ft ↘ ¹/2 mile beyond Much Wenlock.

21 August: 5.30pm 'Race', simultaneous ascent. (a) Harry Salt, amateur balloonist of Tower Place, Shrewsbury, in *Invincible*. (b) Henry Spencer in *Shrewsbury*. No further details, save that Harry Salt reported at Derby in early hours of 22nd!

and rain at several thousand feet! If lost one could always ask the way, as did Henry Spencer on *Shrewsbury V's* maiden flight in 1911. The eight passengers were becalmed for 2¹/4 hours at Berwick, north of Atcham, before descending to pick up 'an angling enthusiast who knew the country'. The evening drawing in they cut their losses and, after admiring the Berth earthworks from the air, landed at Marton Hall convenient to Baschurch railway station.

The most indispensable piece of equipment, however, was *Bradshaw's General Railway and Steam Navigation Guide*. Although the Spencers usually had three, sometimes four, balloons at Shrewsbury, only two were unpacked and inflated at one time. It was highly desirable that where possible Wednesday's balloons should be returned in readiness for Thursday's flights and tethered ascents. With the dense network of railways this was a relatively easy task given the goodwill of country folk to help convey the balloon envelope, neatly packed into its basket, to the nearest railway station and load it into the guard's van.

Some of the distances involved were great, according to both the strength of the prevailing westerly winds and the Spencers' ambition to break the previous year's — or each other's — long distance record. In 1905 Percival Spencer in *Shrewsbury IV*, and with E. R. Cooke, an experienced amateur balloonist, amongst the four passengers, raced the GWR express at over 40mph into Wolverhampton, and continued via Walsall and Birmingham to land eventually at Doe Bank Farm, Great Barr. 1909 was a good year for records. On the Wednesday (18 August) the same pilot, in the aluminised *Condor*, sped at 40-50mph to Tysoe, near Banbury, via Bridgnorth and Stratford. At 16.00 hrs. the next day

Sydney Percival, not to be outdone, took *Shrewsbury IV* some 93 miles to Newark-on-Trent, Notts. His telegram to his brothers at Shrewsbury says it all: 'Splendid trip. Easy descent Newark. Am returning to London. Sydney'. It was handed in at Peterborough on the Great Northern Railway! Indeed, Sydney and his balloon arrived in Kings Cross before the *Condor*, released at 18.30hrs, had returned to Shrewsbury after a relatively short journey to Newport via Wellington. In 1913, in what turned out to be the last race before the outbreak of war, Henry Salt, an amateur balloonist of Tower Place, Shrewsbury, took off in *Invincible* at 17.30 hrs. on Thursday, 21 August, and was last reported over Derby in the early hours of Friday morning!

On the rare occasions that the winds blew southerly or south-easterly a new dimension was added to balloon flights from Shrewsbury – Wales, and everything that implied in the way of mountains, the Welsh language, Welsh place-names, monoglot Welshmen and the eccentricities of Cambrian Railways! On 18 August 1898 the 19.00 hrs. balloon ex-Shrewsbury was put down at Wollaston on the lower slopes of the Breiddens. Percival Spencer could not be persuaded to leave Shropshire and England. Nevertheless both he and his two passengers made a token crossing of the Welsh border, walking along the A458 to Middleton — almost as if they were hoping to get their passports stamped!

The following year, Thursday's (24 August) balloons were both blown northwards. Arthur Spencer's *Victoria* took off at 16.00 hrs. and covered thirty-five miles at a spanking pace to land in a Farndon strawberry field. Newspaper reports draw a tactful veil over damage to crops and the question of compensation, but relations could not have been all that strained as balloon, pilot and passengers were given a lift on the day's convoy of fruit wagons to Broxton station, whence they departed for Whitchurch at 2105hrs, arriving in Shrewsbury at 22.25hrs. These timings would suggest that the balloon party travelled the whole way by market-garden train! Percival Spencer and three passengers followed in *Shrewsbury I* at 19.00hrs. But the wind dropped, forcing them down in a meadow near Bangor-on-Dee station. The balloon was consigned directly to London by the next day's goods, while passengers and pilot wended their way back to Shrewsbury by three separate railway companies via Ellesmere and Whittington High and Low Level Stations, enough to make today's railway buffs drool over their Bradshaws!

Again in 1900, the last day (23 August) saw both balloons blown north-west. The *City of London* (dep. 1600hrs) made Ellesmere via Oswestry in under an hour and came down at Maes-y-llan (Church Field), Erbistock. The *Shrewsbury* (dep. 19.00 hrs.) followed almost the same course but came down at Cross Lanes, St. Martin's. By coincidence both balloons, crews and passengers caught the same train back to Shrewsbury, respectively from Ruabon and Gobowen stations.

By 1906 the Spencers had obviously got over their aversion to crossing the Welsh border. On Thursday, 23 August, Arthur Spencer in *Shrewsbury II*, in thick haze and keeping very low, landed at Llanymynech after a 50-minute flight. At 1900hrs, in *Wulfruna*, Percival Spencer was carried along the disused 'Potts' railway, over the Breiddens and up the Tanat valley, landing at Llanyblodwel conveniently near 'The Horseshoe' pub — 'a peaceful ending to a wild cross-country steeplechase'. There was only one problem. Railway lines in this part of the world did not necessarily mean trains! The Tanat Valley Light Railway only ran to three trains each way on weekdays and the last had left Blodwel Junction for Oswestry twenty minutes before *Wulfruna* had actually left The Quarry! On the Cambrian 'main line' the last Llanymynech-Oswestry train did not provide an onward connection to Shrewsbury. The return journey had to be made by car and lorry, an expensive end to an enjoyable day out!

The response from tenant and landowner to wayward balloonists dropping in unannounced was mixed. Farm workers, once they had overcome their initial apprehension, were generally first on the scene, genuinely concerned and very helpful in securing equipment for salvage and transport. There would always be a few farmers, themselves barely existing above the poverty line, who were more interested in screwing payments 'for damage to crops' from hapless pilot and passengers. Inevitably gentry were courteous, rendering every possible assistance although tending to view balloonists with an air of patronising amusement.

Thus, in 1882, when Whelan and his single passenger, a Dr. Craig, after a 45 minute flight over Sundorne Castle, Haughmond Abbey and Shawbury, landed the *Duke of Edinburgh* in a potato field at Painsbrook, they were met by Sir Vincent Corbett 'and a party of ladies' who, after satisfying their curiosity and examining the balloon at close quarters and sensibly refusing a captive flight, graciously saw to the aeronauts' refreshment and the conveyance of the balloon to Hadnall Station. On the other hand, recalling the flight of Thursday, 22 August 1901, one wonders what would have been the reaction had *Shrewsbury I*, piloted by Arthur Spencer and with Mr. Bernard Head, Ellesmere, on board, actually dragged its basket, anchor cable and flailing grappling hook through the garden party in full swing on the lawns of Lythwood Hall. It was a close run thing with people already heading for cover when the wind lifted the balloon over the house to deposit it gently on the wooded slopes of Lyth Hill nearby.

In 1907 the Wednesday's balloons had mixed fortunes. The 16.30 hrs. flight travelled Atcham-Cressage-Ironbridge-Worfield-Gatacre-Enville Racecourse before coming down at Stourton Fields between Dudley and Stourbridge and missing the last possible train back to Shrewsbury. The 18.30 hrs. flight, with two passengers (including Mr.

A. Poly-Didier of Gresford), came down at Park Farm, Morville, near Bridgnorth. Man and baggage were conveyed back to Shrewsbury by road by no less a personage than Major Atcherley, Chief Constable of Shropshire. The latter's attitude to balloonists had been coloured by his own experiences the previous year when, on Wednesday, 22 August, he had been a VIP passenger aboard *Shrewsbury II* piloted by Percival Spencer. After a 1 1/2 hour flight via Haughmond Hill, Peplow and Tern Hill, the balloon finally came down in the grounds of Stoke Grange, conveniently near Tern Hill railway station, whence the party departed at 1920hrs, arriving in Shrewsbury at 20.18hrs. But how much credit can be given to this fortuitous event being directly responsible for the siting of a First World War aerodrome at Tern Hill, as asserted by some aviation historians, is debatable. The incident occurred two years before the Wright brothers' historic flight, three years before Blériot crossed the Channel and ten years before Tern Hill aerodrome was built!

In 1902 Stanley Spencer had become the first in the UK to build and successfully fly an airship. It had been a modest affair of some 20,000 cu. ft., 75 ft. long 20ft in diameter, with the engine and one-man car dangling from a bamboo girder slung beneath the envelope. It was powered by a 3 1/2 h.p. Simms engine driving a 10ft. wooden propeller. On 22 September it made a one-hour maiden flight from Crystal Palace via Battersea, Hammersmith and Acton to land at Eastcote near Harrow.

This encouraged Spencer to build the larger machine which he brought to Shrewsbury Show on 17/18 August 1906 in addition to the *Shrewsbury* and a back up balloon. The weather on both days was foul. The *Shrewsbury* made just seven captive ascents each day before being tentatively released on prudently short local flights, to Walcot on the Wednesday (carrying three 'proud Salopians') and as far as Donnington on the Thursday.

But the show was stolen by the Spencer airship although it did not fly until the Thursday and then against better judgement, narrowly avoiding disaster. The *Shrewsbury Chronicle*, published on a Friday, could not immediately do full justice to the airship's historic flight and held the full story over for the following week. However, it did give a good technical description of the craft:

> 'But of more popular interest was the great airship with which so many successful flights have been made in the Metropolis and which was now shown for the first time in the rest of England. Messrs. Spencers' airship consists of an elongated gas vessel of torpedo shape, measuring 80ft long and having a diameter of 30 ft., and containing 30,000 ft. of gas. Like the other balloons the airship is inflated by coal gas from the street mains. It carries suspended underneath by Italian hemp chords the keel or framework, a triangular braced structure constructed of bamboo, which contains a car for the aeronauts, the petrol motor etc. and at its forepart the 10 ft. diameter propeller or tractor. At the rear end is the rudder, an immense sail-like construction which is under the control of the aeronauts in the car and by means of this they are able to direct the passage of the airship which attains a nominal speed of 15 m.p.h. Stanley Spencer is in charge'.

Spencer had made four test flights in London before coming to Shropshire. Further sponsored appearances would be made that same year at Hanley and York. The airship was actually capable of holding 40,000 cu. ft. of gas — bad weather may explain the under-inflation — and the use of hydrogen would have doubled its lifting capacity. However, coal gas was already on tap for the other balloons at The Quarry and it cost much less, just over £6 to fill the envelope compared with £140–£150 using hydrogen. The airship now had a 4 1/2 h.p. Simms engine geared down to drive the propeller at 150rpm, making for a cruising speed of 12–15 m.p.h. in calm conditions. Few laymen yet realised that, with such a low-powered engine and the use of coal gas exaggerating envelope size, the airship in practice could make little

The Cardiff-built Willow's airship, 1907.

progress against anything other than the lightest breeze. An alert observer would soon begin to comprehend the realities and problems of the situation!

Preparation of the airship had been completed by early Thursday afternoon, but it was 19.30 hrs. before the strong cross-winds had abated and Miss Gertrude Bacon, daughter of the Rev. John Bacon, took her place in the car and secured her niche in aviation history as the first woman passenger in an airship. The engine was started but it was only 'deft management' that steered the craft clear of all the obstacles in The Quarry. Vertical rise was sluggish and the propeller caught a tent, threatening to capsize the car. As it circled the showground its rudder snagged momentarily in the trees and then, to loud cheers, the airship floated free, 'the first successful airship flight in western England'. It was caught by upper currents, did not have sufficient power to fly contrary to them, and so travelled (or was carried) willy-nilly nine miles due east to land eventually at Admaston, near Wellington, returning to Shrewsbury by road the next day.

A week later, a *Shrewsbury Chronicle's* scoop' (or the editor had his arm twisted!) gave the indomitable Miss Bacon a chance to give a first-hand narrative of events. A rosier-hued picture of the flight duly emerged from her pen:

'The voyage of the Spencer airship from Shrewsbury was, in spite of the unfavourable weather conditions, the most successful airship flight made in this country and goes far to prove that England possesses a dirigible aircraft second to none in efficiency and reliability. Mr. Stanley Spencer was a proud man at the end of his adventure, and I a proud woman, for it had fallen to my lot to be the first lady in England to make an airship voyage

'High winds prevailed over the county on Thursday and prudence dictated the abandonment of the voyage, but 70,000 people were not lightly to be disappointed. The first moment the breeze slackened the airship was made ready. At 7.30 pm we took our places in the basket, the motor was started and the order given to cast loose. As luck would have it, however, a sudden gust swooped down upon us and bore us straight for a large tent besides the starting ground. A shout of horror rose from the multitude, for disaster seemed inevitable, since beyond the tent was a row of lofty limes in which it seemed certain we would be caught.

'It was only the quickness and presence of mind of Mr. Spencer which prevented a catastrophe. In one moment he had dropped an entire bag of ballast onto a vacant spot below. He had disconnected the tractor and stopped the whirling blades, otherwise they would have ripped up the tent and would have brought the rudder sail round to the wind. Crash! went one of the flag sticks on top of the tent and a big spray of lime was borne away on the rudder, but we were clear and in a moment the town of Shrewsbury and the encircling Severn opened out beneath.

'It was immediately evident under what absolute command the airship was. Instantly she answered her helm — bold sweeps and circles to left and right and then the vessel was directed straight against the wind. At our height the current was blowing steadily at about 25 m.p.h., a speed greater than the airship with its present motor can maintain. From this, of course, it follows that we were gradually carried backwards, but the perfect steadiness which she maintained and the strength and power which she exhibited were noteworthy.

'The sensation was at once exhilarating and delightful, more so than in the case of an ordinary balloon. . There was absolutely no pitching or tossing about, only the throbbing of the engine, which by the way, is under the most perfect and ingenious control from the car. At one period Mr. Spence had to quit the car and crawl out onto the framework to adjust the trail rope — a giddy height for such a feat. The rapidly increasing darkness at length put an end to a voyage as delightful as it was unique and a most successful landing was made in a field two miles beyond Wellington, which town we reached about midnight'.

This short flight from The Quarry, Shrewsbury, was the first by a powered flying machine in the north-west. In essence Spencer's airship was little more than a powered balloon but it was a beginning. By September the following year the young Cardiff aviator, Ernest Willows, scarcely nineteen years old, had built and successfully flown his prototype non-rigid airship of some 11,000 cu. ft. but powered by a 30 h.p., eight-cylinder JAP engine — and a new age in British aviation was born, only to fade rather suddenly after 1908 when the country went aeroplane crazy!

Demonstration circuit

Even as North Wales generally, and the Wrexham area in particular, remained untouched by the ballooning phenomenon of earlier centuries, so they were equally slow to embrace the aeroplane. Which ever way one looks at it, the earliest initiative appears to have been taken by Shropshire. Following the formation of the Royal Air Force an inspired *Shrewsbury Chronicle* hack attempted to establish a direct link between the 'super-airmen' of 1918 and one Robert Cadman, the 'primitive birdman' of 1739. The former, as they 'beetled about' over the town, 'displayed a marked infatuation for St. Mary's steeple', the very same tower where Cadman had come to grief some two centuries earlier whilst trying to 'fly' the Severn.

Cadman was a steeplejack who brought a bizarre (and dangerous) act to Shrewsbury, announcing in the local press on 24 January 1739 that 'he intends to fly off St. Mary's steeple, over the River Severn, on Saturday next, flying up and down, firing two pistols and acting several diverting tricks and trades upon the rope, which will be very diverting to the spectators'. A rope was duly slung from the steeple and secured on the Gay Meadow on the east bank of the Severn. The daring 'aviator' wore a breastplate of wood with a groove down the centre to enable him to maintain his own equilibrium as he slid (or 'flew') in the prone position, arms and legs extended outwards. The whole 'flight' or descent of the rope would take about ten seconds and must have been very spectacular as the friction raised a trail of smoke and sparks behind him. Cadman had successfully performed this stunt on three previous occasions, the last at Derby. But 2 February 1739 was not to be his day. The rope broke — too taught for the weather — and the pioneer birdman fell to his death at the Water Lane gate, whilst his wife, imperturbably, but now with an increased sense of urgency, continued to rattle her collecting box amidst the dense crowd of onlookers. Cadman lies buried in St. Mary's churchyard, and on the west side of the church tower a tablet records for posterity:

> Let this small monument record the name
> Of Cadman, and to future times proclaim
> How, by an attempt to fly from this high spire
> Across the Sabrine stream, he did acquire
> His fatal end. 'Twas not for want of skill
> Or Courage, to perform the task, he fell:
> No, no — a faulty cord, being drawn too tight,
> Hurry'd his soul on high to take her flight,
> Which bid the body here below, good night.

Viewed against the endemic patriotism of 1918 it does not seem so strange that a provincial newspaper should latch on to such local associations, however tenuous. It made interesting copy and no doubt was swallowed piecemeal by more credulous readers. However, the Cadman connection unwittingly served to divert the spotlight from a far more legitimate Shropshire claimant to be 'Briton N°1' in the aviation world. In 1903 the Wright brothers made the first sustained powered flight. Over the next few years a number of aviation pioneers, some of whom, like A.V. Roe and the Short Brothers were to become household names, struggled to build their own flying machines. They were chasing the prize of £1,000 offered by the *Daily Mail* to the first man to fly an heavier-than-air machine one mile. Inexplicably overlooked by the aviation world is the contribution made by one Ernest Maund, mechanical genius, owner of a cycle shop in Craven Arms, Shropshire's first motor-cyclist, and probably the builder of the first flying-machine in Britain. In just three months in 1904 he constructed a monoplane to his own design. Built of wood, with a highly polished propeller, it was powered by a second-hand twin motor-cycle engine developing some 10 h.p. The aeroplane was displayed to the public at Craven Arms before a large poster carrying the words 'Briton N°1'. Whether this was a justifiable allusion to

Maund's standing in the field of aircraft construction or was the name given provisionally to his prototype, remains uncertain.

The monoplane was moved out to a large field at Stokesay, where Maund attempted to fly it. It made a good take-off and rose to 'house-top height', about forty feet, before landing. Being rather under-powered it never did succeed in flying the required competition distance. Maund counteracted his disappointment by turning his undoubted talent to motor-vehicles, opening a small factory and garage at Wisemore, Walsall, where, *inter alia*, he invented the first dipping headlight for cars. Ernest Maund must therefore be numbered among the earliest Englishmen, if not actually the first, to fly. The lack of contemporary publicity and his subsequent lapse into relative obscurity is all the more difficult to explain.

Five years later Louis Blériot made his historic crossing of the English Channel, bringing aviation of age. In the same year, 1909, the success of the pioneer Reims Aviation Meeting or air show paved the way for the great aerial sporting events and air displays. These not only made people air-minded but also had a great influence on future developments in aircraft and engine design, especially during the years leading up to the First World War. At 06.30 hrs. on Wednesday, 10 August 1910, Robert Loraine, well known actor-cum-aviator (who had only gained his pilot's certificate the previous June!), took off from Blackpool where he had been participating in a flying meeting. Ninety minutes and 64 miles later, after hugging the North Wales coastline, his Farman racing bi-plane touched down on Rhos-on-Sea golf course. The aeroplane immediately attracted several hundred early-riser holiday makers swelling to thousands as the astute manager of the Llandudno & Colwyn Bay Electric Railway abandoned his summer timetable — already a 10-minute interval service and pressed his reserve tramcars into use to convey the curious to the golf club. The Farman had already been railed off, but little golf was played that day. Thus was North Wales dragged willy-nilly over the threshold of the aviation era.

But prophetic rumblings of great things ahead had been felt by the people of Llangollen a year or so earlier. This was a period of French dominance in the emergent aviation industry. In August 1909 'La Grand Prix de la Champagne', the first of the great international contests, with valuable prizes for speed, high flying and the longest distance, had attracted 38 aircraft and every practising aeroplane pilot in France — but no machine of British design or manufacture. In the first week of October the Paris Aero Show was staged by the Ligue Internationale Aerienne, but again no British planes were there, for the simple reason there was nothing to exhibit. The greatest engineering country in the world had yet to produce a single flying machine. At the two aviation weeks at Blackpool and Doncaster later in October — both unbelievably promoted simultaneously — the large cash prizes on offer not surprisingly found their way into the pockets of French aviators.

In *The Llangollen Advertiser* of Friday, 19 November 1909 that paper's 'War Correspondent' indulges in a wonderful piece of satirical writing berating Llangollen's unpreparedness to meet the threat posed by a rapidly approaching 'fleet' of airships 'plainly of Continental build'. The fact that the 'dispatch' is dated 1 April 1910 [sic] should have alerted readers as to the true nature of the article, but there are savage side-swipes at the Blackpool and Doncaster aviation weeks 'at which the utter inadequacy of British flyers to compete with foreign rivals was demonstrated'.

The writer harked back to 1889/90 when one J. P. Davies of Glynceiriog foresaw 'a time when aerial navies would have to be contended with' and 'a machine. . . that will enable the people from Glyn to fly over to Llangollen in a few minutes. . . and, further than this, will enable you to breakfast on this side of the Atlantic and dine in New York'. Prophetic eloquent words, but strangely uttered in a heated debate surrounding the adoption locally of the Welsh

In 1911 R. A. King flew from Freshfield aerodrome, Southport, to Colwyn Bay. He was the only pioneer aviator to emerge with an enhanced reputation from the weather-hit flying meetings at Rhos-on-Sea in the same year. His Farman 'biplane' is seen on the beach, Saturday, 19 August.

Intermediate Education Act and the formation of the Denbighshire Joint Education Committee!

This surprising piece of satire was inspired, and illustrated, by one of the 'fantasy' post-cards produced by J. Percy Clarke, the Llangollen photographer. It is a contemporary view of Castle Street, Llangollen, looking west towards the bridge. Flying low over the heads of the crowds, having, according to the 'War Correspondent', just dropped a bomb on the public conveniences in Market Street 'completely destroying, in a couple of seconds, work that had taken many years to plan and complete', is an early bi-plane. This is identifiable as a Voisin-Farman, post-May 1908 version, which had given an excellent account of itself in many European countries and had appeared at the Blackpool Air Pageant a month earlier. Clarke must have worked at speed to produce his collage post-card complete with people staring, pointing and shooting at the aeroplane! He would produce a similar collage in 1916 showing German warships on the Dee below Llangollen Bridge being tackled by 'C' Squadron of Llangollen's coracle fleet, while overhead hovered an ancient non-rigid airship identifiable as that built by Lebaudy-Freres of Soissons, France, and purchased by the British government in October 1910. The 'Lebaudy' was more popularly known as the *Morning Post* airship after that newspaper had raised £18,000 towards its cost. It was ill-fated from the start and after a series of embarrassing accidents at Farnborough was scrapped in 1911.

Thus aeroplanes were as yet a novelty, objects of intense public interest that drew large crowds whenever they appeared. The new breed of aviator, backed by enterprising business men, were quick to see commercial possibilities in this phenomenon, particularly the resorts along the North Wales coast with firm sandy beaches for ready-made, renewable landing strips and publicity managers scratching their heads for something new to underpin their summer programmes of attractions and spectacle. However, the flying meetings at Pwllheli (June, 1911) and Rhos-on Sea (July, 1911) were fiascos, but Vivian Hewitt's successful flights from Foryd, Rhyl, were to be a regular feature of that resort's publicity from 1911 to 1914. Hewitt's reputation was further enhanced in April 1912, when he made the first crossing of the Irish Sea from Holyhead to Dublin.

Early seaplane flights were an even greater curiosity than land-based operations. The accompanying illustration shows one Frank Hucks (possibly a member of the extended Hucks aviation dynasty) standing in the cockpit of his Maurice Farman 'pusher' seaplane moored in the Mersey whilst acknowledging the plaudits of the huge crowds lining the landing stage at New brighton some time during the summer of 1914. The aircraft is lettered along the sides 'Frank Hucks Waterplane Co. Ltd.'.

Frank Hucks is probably better known along the south coast as co-founder with another aviation pioneer, Frederick Fowler, of the Eastbourne Aviation Co. in February 1913. In December 1911 Fowler had established the Eastbourne School of Flyning on Willingdon Levels east of the town with a couple of Blériot XIs, teaching himself to fly and gaining his RAeC certificate on 16 January 1912.

The Eastbourne Aviation Co. established a seaplane factory on the crumbles with hangar, turntable and a rail track/slipway down to the HWM. During 1913 the Admiralty became interested and subsidised expansion of the seaplane factory. The School flourished and was visited by many famous aviators. During the summer season Hucks ventured onto the coastal and esturine demonstration circuit, including populous Merseyside on his itinerary. On 6 August 1914, the particular seaplane shown at New Brighton., belonging to the Eastbourne Flying School, was impressed (for £600) by the Admiralty, duly becoming RNAS aircraft serial 887.

Sadly 887's career was less romantic than the high profile and hugely profitable pleasure flying circuit. In September 1914 it was at RNAS Calshot for repair but seldom flew thereafter. It was destroyed on 2 December 1914 when blown over whilst under tow on the Hamble with FSL the Hon. Desmond O'Brien at the controls. The latter only briefly survived the demise of No. 887 and would go missing on 16 February 1915 while on a raid to Zeebrugge. With the outbreak of war Eastbourne aerodrome was requisitioned to become the RNAS's premier *ab initio* training base. The factory was kept busy with repair contracts. Although Fred Fowler took a RNAS/RAF commission, ending the war as a major, Frank Hucks fades into oblivion and is strangely absent from service records.

The Black Country, cradle of the automobile industry and possessing a multiplicity of manufacturing skills and raw materials, quickly turned to the production of aircraft, aero engines and components. Aerial displays and flights, as much an integral a part of publicity, marketing and financial strategies then as now, began to intrude upon the Shropshire rural scene, to a point where any self-respecting *fête champêtre*, be it Wem, Hinstock or Market Drayton, was obliged to mount a display of aerial aerobatics. Aviators were invited, and their

Frank Hucks' Farman 'pusher' seaplane attracts large crowds at New Brighton landing stage.

Benny Hucks' Blackburn Mercury monoplane in its hangar at Ludlow, 1912. [SRRC]

appearances partially subsidised, by the county's forward-looking gentry and landowners, many of whom would be bitten sufficiently badly by the aviation bug as to gain some of the earliest Aviators' Certificates and to run a private plane from their private landing strip.

Flying fever hit Shrewsbury in 1911 with three static displays — a glider, a Blériot monoplane and a Farman biplane — attracting huge crowds at the Shropshire and West Midlands Show. Out in the sticks Shropshire country folk were enthralled by the aerobatics of Bentfield ('Benny') Hucks, of Welsh extraction, in one of the Blackburn Mercury monoplanes, which he had helped test fly from Filey sands. Hucks was a close friend of Ernest Willows, another Welshman, and perhaps the most important British airship pioneer. With his father's encouragement the latter had built his first airship in Cardiff at the splendidly mature age of nineteen! In July 1910 his second airship flew non-stop from Cardiff to London, covering the 140 miles in ten hours, the longest endurance flight at that time by a British airship. His third, the *City of Cardiff*, made the first London-Paris flight by a British subject in November of the same year. In 1911 he would establish the firm of E. T. Willows Ltd. in Birmingham, stabling and testing his airships and balloons at Castle Bromwich. Not surprisingly, neither Hucks nor Willows are known to the *Dictionary of Welsh Biography*, not being preachers, composers of hymn tunes or of a literary bent!

In 1911 Hucks was giving passenger flights from the Midlands Aero

Club's ground at Castle Bromwich and on 10 September became the first man to fly the Bristol Channel, from Weston-super-Mare to Cardiff. In 1912, as his fame spread, he was sponsored by the *Daily Mail* to give flying displays throughout the Midlands. In Shropshire he gave demonstration flights from Sych Field, Market Drayton, and from Ludlow. It is in this county and along the Welsh border that his operations overlapped with those of his friends and great rivals, the 29 year-old Irishman, John ('Jack') Brereton, and Gustav Hamel, an Englishman of German extraction, son of the physician to Edward VII.

'Jack' Brereton was born in Galway on 22 July, 1882, and learnt to fly on a Bristol 'Boxkite' biplane at the Bristol School of Flying, Brooklands, gaining the Royal Aero Club Aviator's Certificate No.136 on 19 September 1911. He succeeded Hubert Oxley (killed in a flying accident on 6 December 1911) as test and demonstration pilot for the Blackburn Aeroplane & Motor Co. (of Leeds) at Filey aerodrome. He also doubled up as a flying instructor at the Blackburn Flying School, Filey. The latter would move to Hendon in September 1912. During the war the company would produce the Blackburn 'Baby' Seaplane and the Blackburn 'Kangaroo' twin-tractor biplane, a long-distance bomber much used in anti-submarine patrols over the North Sea and elswhere. But for the moment Robert Blackburn was devoting his energies to perfecting a military version of his popular Mercury III monoplane.

Early in May 1912, at Filey, Brereton test flew a Mercury III with the 50hp Isaacson engine, produced by the Isaacson Engine Co., also at Leeds. It was an uncowled stationary radial but had considerable teething problems, causing two forced landings (at Malton and Welham Park) on a flight from Filey to Leeds on 29 May, necessitating an undignified return to Filey by rail!

A Mercury III monoplane, this time powered by a 50 h.p. Gnome, was test flown at Filey on 7 June 1912, and it was this version that was flown by Brereton as a demonstration aeroplane at shows and fetes at Wem, Shrewsbury, Newtown and elsewhere. From contemporary publicity postcards (with inset medallion of the pilot) the machine was immediately identifiable by its half-cowled engine and having Blackburn emblazoned on the fuselage.

Strangely, after the Hendon school closed in June 1913, Brereton is lost in obscurity. The war put an end to flying displays in Shropshire and along the Welsh border, but there is no trace of him in the RFC or RNAS. He may have become a test pilot for another company, but details of such posts are difficult to come by.

Educated at Westminster School and a graduate of Cambridge University, the 23 year-old Gustav Hamel had looked for something novel and worthwhile in life and, caught up in the spirit of the age, embraced flying. He took lessons at Isay-les-Moulineaux (Seine), presumably at the René Caudron Flying School, and gained his Aviator's Certificate in January 1911 and RAeC Certificate N°64 on 14 February 1911 in a Grahame-White Blériot monoplane. He became chief instructor at the Blériot School, Hendon and in March 1911 won the inaugural Hendon–Brooklands race. Hamel was by far the most accomplished aviator to visit North Wales, and the first to fly from Wrexham. Flying a Blériot monoplane (1911 price, £475 new!) which could attain the great speed of 65 m.p.h., he had already crossed the Channel eleven times (twice with a lady passenger) and made the journey from London to Paris in a day. Flying a Bleriot for the Grahame-White Aviation Company, Hamel had inaugurated the regular 'aerial post' (9–26 September 1911) from London (Hendon) to Windsor on the occasion of the coronation of George V. A Farman was also used, and a total of 25,000 letters and 90,000 postcards were carried. Grahame-White had earlier (10 August 1911) made the first unofficial airmail flight from Blackpool. Hamel had won aerial races at Sunderland and Sheffield and in 1913 was to win the Aerial Derby at Hendon (a race round London for prize money and the *Daily Mail* Trophy) piloting a Morane-Saulnier monoplane. More recently he had broken the world altitude record standing at 11,500ft.

On 13 April 1912 the Royal Flying Corps came into being, with Naval and Military Wings. In early August, at Larkhill on Salisbury Plain, the Army held trials to select the aeroplane best suited to its needs. The minimum specification was for a machine that could lift 350lbs in addition to instruments and fuel and oil for 4½ hours, that could maintain an altitude of 4,500ft for an hour and reach 55 m.p.h., that could operate from ploughed land, uncut grass and clover, and that would meet certain criteria relating to ease of handling, noise levels etc. Only twenty-two of the thirty two invited aircraft constructors turned up on the day, but these included a Blériot 'Sociable' monoplane piloted by Gustav Hamel.

Hamel's brief but profitable association with Shropshire and North Wales began by default — on the occasion of a Flower Show at Mount Pleasant Farm, Hinstock, near Market Drayton. Looking ahead to 1912, the organising committee had pencilled in one of the new-fangled, crowd-drawing flying displays. It was scheduled to be given by the Hon. C. S. Rolls (of Rolls Royce fame), one of the rich adventurers who turned to flying as their contemporaries took to car racing. Charles Rolls had been a respected name in aviation circles for some time, but more as a balloonist. Now, as the proud possessor of Aviator's Certificate No. 2, he had bought a Short-Wright biplane and set about establishing records. On 2 June 1910 he had become the first man to fly across the English Channel both ways. On 27 June he

Gustav Hamel in the cockpit of his Blériot.

participated in the first 'All-British Flying Meeting' at Dunstall Park, Wolverhampton, making his way immediately afterwards to another aviation meet at Bournemouth, where tragedy struck. On 12 July his Short-Wright broke up in the air and Rolls fell 200ft to his death, involuntarily setting another record — as the first British aviator to die. At the invitation of the squire, Capt. the Hon. Gerald Clegg-Hill, Hinstock Villa, Rolls's appointed place at the Hinstock Show on Thursday, 18 July 1912, was taken by Hamel in his Blériot monoplane. He was a great success.

Because of the cramped nature of Hinstock's usual showground the venue had been switched to Mount Pleasant Farm where Arthur Butler had put a 14- and 20-acre pasture field at Hamel's disposal. The latter had been at Hull on the Monday. He was at Loughborough on the Tuesday but arrived late the same evening at Market Drayton station, complete with aeroplane! His Blériot monoplane was housed in Mr. Butler's cart-shed for the night. The next day the canvas hangar was erected, the wings put on the machine, and the mechanics set to work. But it was late on Thursday before weather and aeroplane were declared fit for flying. The patient crowd was rewarded at 3.15 pm by a 4-minute flight at 200ft around the showground. At 6.15 pm there were the usual thrilling aerobatics and a couple of circuits of Market Drayton at 500ft before the climax — a climb to 2,000ft when the engine was switched off for the long heart-stopping glide down to a perfect landing.

On 27 August Hamel was at Macclesfield and at Buxton on 5 September, whence he flew on to Whitchurch. On Wednesday, 16 October, he was at the Chester City football ground, Sealand Road. After a civic reception, Hamel gave three demonstration flights, at 1430, 1530 and 1630hrs, part proceeds going to the Chester Royal Infirmary. On 19 October he was flying from Orleton Park, Wellington and appeared at Stoke on 24 October. Thus were laid the foundations of a very lucrative circuit, at least one town, often two, a week in season, the more prestigious air races and flying meetings permitting. By present-day standards the 'flying season' was over-long, extending well into late autumn. Indeed, the *Whitchurch Herald* informed its concerned readers that on Saturday, 9 November 1912, whilst demonstrating at Burnley a gust of wind had blown Hamel's Blériot against some telephone wires, damaging the propeller and giving the popular aviator 'the narrowest escape he had ever had'.

Hamel's visit to Whitchurch on Wednesday, 12 September 1912, was the star attraction in that year's 'Cycle Parade and Floral Carnival'. But the event was badly organised. The flying took place between 14.30 and 17.00 hrs. on the large fields of Belton Farm, a mile out of town

Hamel's Blériot XI at Hinstock Show, July 1912. The same machine (Nº 3) was flown at Chester, Whitchurch and Wrexham later in the year. [SRRC]

on the Wrexham road, where also would take place from 14.30 hrs. onwards the 'Grand Pageant of Dancing Troupes'! But the viewing and judging of the cycles was hived off to the Smithfield near the town centre. Nevertheless the whole day made a small profit, the proceeds being shared by the Cottage Hospital, the Arts and Crafts Guild and the provision of seats in Jubilee Park. Because of the split sites the *Herald* leader writer felt justified in castigating the organisers 'as to the enormous number of people who seemed to have set out on Wednesday last with the determination of getting as much as they could for nothing. We dropped a hint about this last week; and it was not a little surprising to see literally hundreds of motor car owners and others, obviously well able to go onto the field, lining the roads and fields in various parts and thus getting a view of the flying for nothing. To say at least it was a rather disreputable thing to do, and in calling attention to it one cannot help regretting that so large a number of people should have been found capable of doing it'!

It was the first aerial display in the Whitchurch area. Great expense was involved especially in view of the uncertain weather in a 'catchy season'. As it was there was a bitingly cold north wind necessitating a short trial flight at 13.00hrs. Even so there were over 5,000 people present, so the two demonstration flights had to go ahead although with little scope for 'fancy flying'. But human nature being what it is the *Herald's* criticisms fell on deaf ears! It cost to go on the ground. The Blériot was hidden from view in its canvas hangar in another field 150-200yards away from the crowd. No one was let into its enclosure during pre-flight checks, and even between flights there was 'a small charge of 6d. ' for those wanting to view the aircraft at close quarters.

For the next twenty-seven years, following its rather acerbic coverage of Hamel's visit to the town, the *Whitchurch Herald* would continue to comment on aviation matters, both at regional and national level. Two weeks later, in its issue of 28 September 1912, it ran the story of the death of H. J. D. Astley, killed on 24 September at a Belfast flying meeting as he desperately tried to avoid the crowds on a rather restricted flying ground. Astley is better known to local historians as the billed star (inexperienced but cheap!) for what was heralded as 'The First Flying Meeting in Wales' scheduled for Whit week, 5/6 June 1911, at Pwllheli. In the event Astley crashed his 'Birdling' monoplane on the first attempted take-off and that was that! In July he had lined up with twenty others at Brooklands for the start of the 1,010 mile *Daily Mail* 'Circuit of Britain'. but failed to complete the course. Strangely his stature remained undiminished. But the human interest prompting the *Herald* to cover his death lay not so much in his non-achievements as in the fact that Mrs. Astley had been prominent in Hamel's 'retinue' during his visit to Whitchurch a fortnight earlier, giving rise to considerable speculation and unkind gossip!

On 31 October, at the invitation of Lord Harlech and in aid of the county's sanatorium scheme, Hamel was at Oswestry. Despite strong winds and occasional rain showers, he managed to give two remarkable aerial demonstrations, before a paying crowd of some 3,000, from a somewhat restricted flying ground at Brogyntyn Park. He made several circuits of the town, enabling collecting boxes to be taken into the streets and hostelries!

On Sunday, 6 October 1912, Hamel literally dropped in at Wrexham, landing in a large field at Whitegate Farm, Hightown. He had covered the 26 miles from Crewe in 20 minutes, pursued by his groundstaff and mechanic in a racing car. Hamel had landed to ask the way. On the Monday, after being liberally entertained in the Officers' Mess at the Barracks, he took off again for the Racecourse on the other side of town, where exhibitions of flying were scheduled for the Wednesday, and where a hangar had already been erected.

Hamel was flying a modified Blériot XI 'tractor' (as distinct from 'pusher') monoplane. With the new 50 h.p. Gnome rotary engine designed by the Seguin brothers, a Chauviere two-blade propeller, and other alterations to elevators and lateral controls, the Blériot XI had assumed the definitive configuration that proved to be the making of Louis Blériot as a designer and aviator. It was an outstanding aircraft. Apart from its sporting achievements, it was the first aircraft to be sold to the French Army and the first to be used in war (Italo-Turkish, 1911).

At the Racecourse, home of Wrexham AFC and traditional venue for public displays and military parades, the landing strip was just beyond the white running rails of the course itself, still in situ. What was to turn out to be the last of the Wrexham Races had been held on Saturday, 31 August 1912.

The organiser of this and other Hamel exhibition flights in North Wales was one C. E. Hickman, a Wrexham business man. His souvenir programmes informed the inhabitants of Wrexham that 'Your town is fortunate in the opportunity of being able to witness such an exhibition as Mr. Hamel always gives to his audience. His bankings, vol-planes [glides] and pancake descents are extremely thrilling. But these fascinating intricacies. along with the beauty and grace of flying can be seen only from the Flying Ground, certainly not from the street or road side'. Admission to to the special enclosures and to the old racing grandstand was 1/- and 6d. Clearly Hickman had learnt something from the Whitchurch fiasco and was out to maximise takings and his promoter's cut. Special trams had been laid on from Rhos to the 'Turf Tavern' terminus at the Racecourse, and the former paddock was roped off to provide a privileged enclosure for the motor cars of gentry. By Hamel's arrival at 14.15 hrs. over 5,000 spectators were on the ground.

Demonstration flights were scheduled for 1440, 1530 and 1630hrs, but such was the press of people eager to inspect the Blériot XI that the first flight was delayed by nearly an hour. The local newspaper takes up the report: '... fortunately the weather was perfectly calm and admirably suited for aviation, with the result that Mr. Hamel was enabled to demonstrate to the full his remarkable skill and the possibilities of the monoplane. The second performance was notable for the high altitude reached, about 3,000ft — and for a daring flight over the town. Darting away from the course in the direction of Cobden Mills, the machine sped at tremendous speed over Newtown and Hirdir [present Huntroyde area] and returned over the Parish Church tower and Rhosddu to the field, where further graceful evolutions were witnessed'.

'In the third flight the aviator kept fairly low, and with the exception of a sudden dash around the chimney of Messrs. Powell Bros. & Whittaker's foundry [alongside Wrexham General railway station] did not depart from the Racecourse. It was the most nerve-racking performance of the afternoon, however. The engine was several times stopped, but when the machine was within a foot or so of the ground the engine was restarted and the next moment the bird-like structure was soaring away hundreds of feet in the air. It was enthralling and hypnotising, and those fortunate enough to witness it will not quickly forget the wonderful exhibition of aircraft and the remarkable skill of the aviator... '

On Thursday, 7 November 1912, Hamel visited Shrewsbury, under the auspices of one G. Tupling, another would-be entrepreneur of Whitchurch. The display had been widely advertised and for miles around the event was marked as a quasi-public holiday. Over 6,000 people paid for admission; school children were admitted free. The flying took place from a large field adjacent to the old Shrewsbury Racecourse in Monkmoor. and this possibly goes some way to account for the location of a small aerodrome here in 1917, vestiges of which would survive to become a Second World War RAF Maintenance Unit. Hamel was afforded a civic reception and was presented with a specially struck Gold Medal. Conditions were ideal save for some strong winds. Hamel made the traditional three ascents, each of increasing difficulty. It was on the first that he earned the headlines 'Thrilling Moment! Aviator's Narrow Escape!' As a local reporter had it: 'Either Mr. Hamel struck a gust of wind or he made a miscalculation, with the result that only by superb skill and presence of mind was he saved from falling to earth'. He had actually hit an air pocket. The incident underscores how, very much against their better judgement, these early aviators were forced to take to the air when faced with restless crowds who did not appreciate the perils of flying in treacherous weather conditions.

Hamel revisited Wrexham Racecourse in 1913 when he was advertised to give exhibition flights between 15.00–17.30 hrs. on Wednesday, 11 June, and Saturday, 13 June, with public viewing sessions on 10–14 June. According to the advanced publicity, since his last appearance at Wrexham Hamel had gained further renown by successfully making 'a wonderful journey from Dover to Cologne'. But this time events at the Racecourse did not go without mishap, as a contemporary newspaper account reveals:

'... Mr. Hamel, who had a cordial reception, went up about 3.30 pmand after three circular flights in which he at times attained a great speed, alighted on the playing pitch close to the town goal. He ran along at a rapid rate towards the other end of the field, and owing principally to the lack of grip because of the wet grass, was unable to avoid coming into contact with the railings, with the result, although the impact was not great, one of the main upright tubes of the fore-carriage was buckled. This, unfortunately made further flying impossible and the machine was moved back to the tent.

'Although the flight was devoid of thrills it was followed with much interest and showed the mastery of the famous aviator had over the machine. He was naturally disappointed. at the unfortunate accident, and had intended going up again immediately and treating the spectators to some of his thrilling banking, vol-planes and pancake

descents, as the conditions, he said, were by no means unfavourable, though the breeze was fairly fresh. The machine, a Blériot monoplane fitted with a 50 h.p. Gnome engine, was quite new and this was the only the second occasion on which it had been used. Its graceful lines and general appearance were greatly admired.

'Mr. Hamel at once wired to London for the necessary materials to repair the damage and all will be in readiness for the flights announced for tomorrow (Saturday) afternoon. Where necessary, the fencing on the Racecourse will be temporarily removed to avoid a repetition of Wednesday's accident. Mr. Hamel, who afterwards left by car for Shrewsbury, from whence he took train to London, is expected to fly from Hendon to Wrexham with Miss Trehawke Davies, his companion in many daring flights, today (Friday), It is also announced that anyone desirous of ascending will be taken up on the payment of a fee'.

But even this was not to be. Illness intervened, with Hamel finally giving his exhibition flights on Saturday, 28 June 1913, en route to the Welsh National Agricultural Show, Porthmadog. The fact that the Blériot XI was only on its second outing would seem to suggest that the damage suffered by Hamel's previous machine during an ill-fated demonstration at Penrhyn Castle, Bangor, on 12 March 1913, had been more serious than contemporary accounts would suggest and had been written off. On that occasion, after some delay and against his better judgement, with winds freshening ominously, he had made a fine take-off, but at the height of 100ft was caught by a strong gust of wind, wavered momentarily, and was tossed to the ground amidst cries of alarm from onlookers. Only his superlative flying skills enabled Hamel to lessen the impact and to escape with just minor cuts to the legs.

Hamel's companion, Miss Trehawke Davies, was quite a character, a 'liberated' lady ahead of her time. In September 1912 the local press dwelt at length on Hamel's arrival at the improvised aerodrome at Belton, near Whitchurch '...in a motor car accompanied by several lady friends...' — such was the impact made on the fairer sex by 'this stripling, barely 23, as frail in build as his machine, yet who bore himself with an air of nonchalance... '

The redoubtable Miss Davies, however, must be ranked somewhat above these fragile Edwardian flowers. The Wrexham souvenir programme confides that she was the brave young lady who had accompanied Hamel on several of his cross-Channel trips and hints that she was an aviator in her own right as well as being a writer on matters aeronautical for the tabloid press of the day, notably the *Penny Pictorial*! In the former it was wrong. Despite her craze for flying, and owning two Blériot monoplanes, Trehawke Davies never took flying lessons, never flew as a pilot, and never qualified for her Aviator's Certificate. She was wealthy enough to employ the like of James Valentine, Bentfield Hucks and Gustav Hamel to fly her around the United Kingdom and the Continent. It was with Hamel in a Morane-Saulnier, at 1,000ft above Hendon on 2 January 1914, that Miss Davies gained the distinction of being the first woman in the world to loop the loop, followed by a half roll for good measure. The first Englishman (Welshman!) to loop the loop was Benny Hucks on 13 November 1913, he having travelled to Buc (France) to learn the manoeuvre and other aerobatics from the French aviator Adolphe Pegoud.

On 28 August–3 September 1912, barely four weeks before their Wrexham appearance, Miss Davies had been the passenger in Thomas Sopwith's Blériot two seater monoplane, piloted by H. J. D. Frankland-Russel-Astley on an arduous seven day cross-country marathon Hendon–Hardelot-Isay-les-Molineaux–Berlin–Mezieres–Bonn. As if this did not supply headlines enough, she and her pilot were fortunate enough to survive the crash of the Blériot at Lille the following Wednesday, 4 September. The pilot lost control when the heel of a flying boot went through the cockpit floor, momentarily preventing the use of the rudder bar! The machine side slipped from 200ft into the ground. No wonder Wrexham people were agog to see this lady in the flesh!

His 1913 flying display would prove to be Hamel's last visit to Wrexham. In September 1913, flying a Morane-Saulnier to replace his temporarily damaged Blériot, he won the Walsall Air Race between him and Benny Hucks, sponsored by the *Birmingham Daily Post* which offered a prize of £500 and a trophy to the winner. In May 1914 Hamel undertook a non-stop flight (340 miles in 4 hrs. 14 mins.) to Germany, but was lost on the return journey, presumably coming down in the North Sea. At the time he was also preparing for an attempted flight from Ireland to St. John's, Newfoundland, for the prize of £10,000 offered by the *Daily Mail*. Hamel was the projected pilot for the Martin-Handasyde Transatlantic Monoplane, a two seater with a 66ft wingspan and a new Sunbeam Mohawk 225 h.p. engine. The outbreak of First World War saw the end of such attempts. From being just a sport and an adventure, flying suddenly became a very serious business indeed.

As at Monkmoor, Shrewsbury, Gustav Hamel's choice of the old Racecourse at Wrexham as an improvised airfield in 1912 and 1913, established a precedent. Its use, actual or proposed, forms a major thread in any narrative of the history of flying in the Wrexham area. Because it long survived as an open space, as playing fields and a recreational area, the Plas Coch/Ashfields suburb of Wrexham would re-emerge in the 1930s as the preferred site for a proposed municipal aerodrome and, in the 1950s, as the intermediate picking-up and dropping-off point on the short lived BEA experimental Liverpool–Cardiff helicopter service.

The Bristol-Coanda monoplane (No 14) at the Military Aeroplane Competition, Larkhill, August 1912. This was the machine in which Lt. Bettington and 2/Lt. Hotchkiss were killed in September. [SRRC]

It is interesting to reflect that, despite the powerful advocacy of Hamel, Hucks, Brereton and colleagues on the flying display circuit, monoplanes formed only a small proportion (17%) of the first 200 assorted aircraft taken on strength by the newly formed RFC (Military Wing) in 1912–13. Of its 34 monoplanes the Blériot XI-2 (9), XI (5) and XXI (1) led the field, followed by the Deperdussin (7), Nieuport IV.G (5). Bristol-Prier (2), Bristol-Coanda (2), Flanders F4 (2) and a single Martin-Handasyde.

At this point in the story of military aviation the local historian's ought to note the extent to which local individual philanthropy furthered the cause of aircraft research and development, sadly neglected and under-funded at the Royal Aircraft Factory. In June 1910 Hugh Richard Arthur Grosvenor, 2nd duke of Westminster with his seat at Eaton Hall between Chester and Wrexham, had jointly presented (with Col J. Laycock of the ENV engine company) a Blériot XII monoplane to the War Office. It duly arrived at the RAF in a rather dilapidated state after some rough handling at the Royal Engineers Balloon School. Under the guise (for accounting purposes) of 'repairing' the machine, the RAF completely reconstructed and modified the Blériot to produce a totally new machine, Geoffrey de Havilland's small pusher 'canard' biplane, the S.E.1, retaining only the 60 h.p. ENV engine. This particular venture came to a premature end when the S.E.1 crashed on an ill-advised test flight on 18 August 1911.

But by this time de Havilland was interesting himself in the potential of the propeller-first tractor-biplane. Again the first BE1 was another devious reconstruction job, this time on an old Voisin pusher, again presented to the War Office in May 1911 by the duke of Westminster and transferred to Farnborough for 'repair' the following July. Only the Voisin's 60 h.p. Wolseley engine was incorporated into the new two-seater tractor machine, and even this was soon replaced by a 60 h.p. Renault engine. Geoffrey de Havilland first flew the BE1 on 4 December 1911. It was handed over to the Air Battalion in March 1912. As aircraft No. 201 (RFC Military Wing) it had an illustrious career with Nos. 2 and 4 Squadrons at Farnborough and Larkhill, being used latterly in wireless experiments. Re-engined with an 80 h.p. Renault it was still in use at the Central Flying School in June 1916.

The duke of Westminster was no mere dilettante in things mechanical. He was a founder member of the Brooklands Automobile Racing Club and 1914–16 commanded an armoured car detachment (Royal Naval Division). As patron of the living of Bangor-on-Dee, near Wrexham, the duke's military exploits were followed with great interest. The *Border Churchman* for June 1916 chronicles his award of the D.S.O. and the whole parish basked in the reflected glory. The duke had started the war as a Temporary Commander RNVR with armoured cars in France. It was a pioneer unit and the duke was mentioned in dispatches. In 1916 he led his squadron, by then a regular military unit, across the Libyan Desert to rescue sixty survivors of the merchant ship *Tara*, which had been torpedoed in the Gulf of Salum (east of Tobruk), and who had been taken prisoner by Senussi tribesmen. The Easter Vestry formally resolved to record the duke's exploits in its minutes and to send a letter of congratulations to a patron 'who had always readily and generously supported any scheme for the well being of the parishioners' from a PCC that 'had every reason to be especially grateful to the Duke's family for the succession of excellent rectors appointed to the benefice. . . '

The duke's interest in aircraft may have prompted his nephew, Capt. Robert ('Robin') Arthur Grosvenor (1895–1953), son of Lord Arthur Grosvenor of Broxton Hall, near Farndon, to transfer to the RFC in 1916 and to become an 'ace' fighter pilot. Educated at Wellington College, Robert Grosvenor had been commissioned into the 2nd Dragoon Guards, earning an M.C. in 1915. He transferred to the RFC in October 1916, flying two-seater 'pusher' FE2ds with 57 Squadron out of St. André-aux Bois and Flenvillers before converting to single-seaters and being posted to 84 Squadron equipped with S.E.5As. The latter unit shifted quickly from aerodrome to aerodrome — Flez, Champien, Vert Galand, Conteville and Bertangles — during which six months Capt. Grosvenor claimed eight enemy aircraft destroyed (one shared) and seven 'probables' (one shared), that is, enemy aircraft last seen falling to earth 'out of control'. He was mentioned in dispatches and in 1918 collected a bar to his MC.

Eleven of his sixteen claims were made flying what was obviously his 'personal' mount, S.E.5a serial B8408. This aeroplane had been taken on squadron strength on 28 March 1918 and Grosvenor's first kills flying it came on 3 April when he downed two Pfalz DIIIs within one minute. Some twenty-seven aircraft of 65 and 84 Squadrons had tangled in an hour-long dog-fight with thirty enemy aircraft., destroying five of the latter. This so impressed Maréchal Foch that large formations of fighters on offensive patrols henceforth became the norm as units strove to meet his directive to 'seek out and destroy enemy aircraft'. Grosvenor last flew B8408 on 18 May when he shot down an Albatros DV and a DFW C just before lunch. Shortly afterwards he was posted to Home Establishment. Sadly, on 27 June, B8408, now piloted by a South African, 2/Lt. C. R. Thompson, was badly shot up in combat with Ltn. U. Neckel of *Jasta 12*, and was further damaged in a heavy landing back at Bertangles. It was struck off squadron books and two days later ended up ignominiously on No. 2 Salvage Dump. Cecil Robert ('Ruggles') Thompson, after this rude introduction to aerial warfare, would go on to become an 84 Squadron 'ace' in his own right, claiming six enemy aircraft and balloons and winning the DFC in the process.

RFC establishment figures were not reached until the outbreak of the First World War, and then only by calling up civilian pilots and appropriating their best machines for the Central Flying School. At the beginning of 1913 the RFC possessed only thirteen machines in flying order, a situation exacerbated by the ban on monoplanes for military purposes imposed 14 September 1912. This would set the cause of British aircraft design back nearly twenty years. 1918-36 was the 'Golden Age' of the biplane, with technical progress so slow that it makes the advances and achievements in design and performance between 1936 and the outbreak of the Second World War all the more remarkable and impressive.

In civilian flying circles the monoplane continued to perform well. In October 1913, for example, an 80 h.p. version of the Blackburn Mercury won the Inter County 100miles 'War of the Roses' Air Race, sponsored by the *Yorkshire Evening Post*. Additionally the Mercury III had been very popular with officers of the RNAS.

This notwithstanding, the cause of the ban was the growing suspicion in official circles that two wings might be better than one, a gut reaction fuelled by the death of Capt. E. B. Loraine and his passenger in a Nieuport on 5 July, followed by two accidents in quick succession during the Army manoeuvres in August/September 1912. The first to come to grief was a two-seater 100 h.p. Deperdussin monoplane (serial 258) of 3 Squadron. On reconnaissance duty with the cavalry in Hertfordshire, it broke up over Gravely on 6 September 1912, killing its two crew, Capt. Patrick Hamilton RFC, ex-Worcestershire Regiment and Air Battalion, and Lt. Athole Wyness-Stuart RFC, ex-RFA Reserve. Four days later, on 10 September 1912, a Bristol-Coanda monoplane (serial 263) of 3 Squadron, also crashed, killing Lt. C. A. Bettington, observer, and 2/Lt. E. N. Hotchkiss, Special Reserve RFC, pilot.

Two days later the order banning the flying of all monoplanes by the Military Wing was promulgated, seemingly justified by the crash on 15 December at Wembley of a Handley Page Type F monoplane, killing Lt. W. Parke and his passenger Alfred Arkell Hardwick, assistant manager to Handley Page.

Claude Albermarle Bettington was born in Cape Colony and had been a Royal Artillery officer during the Boer War, surviving the siege of Ladysmith. He had subsequently left the service but had been recommissioned in the RFC on formation in April 1912, gaining RAeC certificate No. 256 in a Bristol bi-plane on 24 July 1912. At his death Devonshire-born Lt. Parke was easily the most experienced naval pilot, gaining RAeC Certificate No. 73 on 25 April 1911. There followed a period as an Avro test pilot before he was commissioned into the RFC Naval Wing and posted to RNAS Eastchurch, test flying Short seaplanes. He regularly spent his week-end leaves at Brooklands or Hendon, acquainting himself with other manufacturers' latest designs. This was how he came to be flying the H. P. Type F in which he was killed.

The accident to Coanda No. 263 was given wide coverage in the local press, not because of the far reaching implications for aviation history, but because 2/Lt. Hotchkiss was a Shropshire man, a native of Craven Arms but even better known in the Oswestry area where for seven years he had been brewer to Messrs. Dorsett, Owen & Co. (swallowed up by Wrexham-based Border Breweries in 1931). He played hockey regularly for Oswestry Men's XI, gaining several county caps. He gained RAeC Certificate No. 87 on 16 may 1911 and moved to the British & Colonial Aeroplane Co. , Ltd. , Filton House, Bristol, the first great British aircraft manufacturer, founded in 1910. Latterly Hotchkiss had been Chief Instructor at their School of Aviation at Brooklands. Amongst his pupils had been the unfortunate Bettington.

The wreckage of the ill-fated Bristol-Coanda monoplane (bearing the number 263), which crashed at Port Meadow, Oxford, 10 September 1912. [SRRC]

At 07.03 hrs. on that fateful day they had set out in No. 263 from Salisbury Plain on a cross-country exercise. As the first staging-post was appproached, control was lost. At 08.15 hrs. the last entry in the log read: 'Over Oxford. Struck rainstorm. Very wet and uncomfortably windy'. They came down from 2,000ft to 600 ft., intending to land on Port Meadow but overshot and crossed the river. As the plane turned there was an explosion, a loud bang and cracking sounds. The left wing buckled and the Coanda fell to earth, the tail unit in the stream, the rest crumpled up amidst the willows on the river bank. Hotchkiss was dead, with a fractured skull. Rescuers did not realise that there had been a second crew member until the shapeless mass of clothing seen floating in the water was pulled ashore. Bettington had been thrown clear but had died instantly from a broken neck. No. 263 with its 80 h.p. Gnome engine had been on Army books a fortnight. After tests at Parkhill Camp (Military Trials No. 14) it had been purchased by the War Office, since when it had had its small rudder replaced by a larger one. A RAeC enquiry blamed the failure or the accidental release of a quick release catch attached to a steel strap anchoring the flying wires on the bottom of the fuselage, producing strains which tore them away.

Port Meadow was one of those broad open spaces that, in an era of very tentative, literally 'short-haul' cross-country flying, had been pencilled in by early aviators as a suitable landing ground, whether permission had been given or not. Hubert Latham, English sportsman resident in France, graduate of Balliol, and pilot/designer for the Antoinette Company, was probably the first to land an aircraft there, an Antoinette IV on 19 May 1911 after a flight from Brooklands. The Army had scheduled tests for the ground for 14 June 1911, but the 50 h.p. Gnome Henry Farman biplane (serial F1) earmarked for the task could not cope with the strong winds, finally landing at Bessels Leigh polo ground five miles to south-west. A second attempt the following day also had be abandoned as the Farman was forced down by near gale force winds at Heathercroft Farms, Abingdon as it struggled to return to Salisbury Plain.

By November 1911 the Imperial Aero Club were on site. It lost fifteen aircraft., valued at over £8,000 when its aircraft shed was demolished in the first of the winter's gales. Private flying from Port Meadow suffered a further blow when, on 5 August 1913, five aircraft sheds were destroyed by fire, thus paving the way for the military to move in. Army and Naval aircraft had used the field throughout, and during the September 1913 manoeuvres 3 Squadron, Netheravon, was based there with its motley collection of Farman F20s and Blériot XIs. Port Meadow was also used for general military training as well as an airfield — the Army's *Beta II* airship called there on Easter Monday, 24 March 1913. For aviation purposes the field was taken under the aegis of 21 (Training) Wing established at Cirencester in August 1916.

The first wartime unit to arrive on site was 40 Reserve Squadron, raised at Northolt in July 1916 and flying in to Port Meadow on 21 August 1916. On 31 May 1917 all Reserve Squadrons were designated Training Squadrons (TS). On 1 June 1917 No. 17 TS flew in from Waddon (Croydon) but did not stay long, moving on to Yatesbury (Wiltshire) on 8 October. They were replaced on 10 October by No. 1 TS from Narborough. The latter, along with 40 TS, took their Camels and Pups to Beaulieu in January 1918. On 16 December 1917 35 TS was posted in from Northolt, under the command of Major Lockett Henderson. Also in December 1917 34 TS arrived from Castle Bromwich, followed by 71 TS from Netheravon on 30 March 1918. In August 1918 both were amalgamated to form 44 TS.

As a relief landing ground Port Meadow received numerous lame ducks but none as spectacular as the giant Handley Page 0/400 bomber (serial D5401), piloted by Lt. Shaw, which dropped in on 19 April 1918. Its twin 375 h.p. Eagle VIII engines were having fuel injection problems on the delivery flight from the builders in Birmingham to Lympne. This took a remarkable nine days, with enforced stops at Upper Heyford and Farnborough to try and clear the problem. The RAF had relinquished Port Meadow by 18 February 1919 but the area continued to be used spasmodically in emergencies. Thus on 4 August 1926 one notes a Gloster Grebe II (J7572) of 25 Squadron (Hawkinge) suffering engine failure on take-off after an earlier forced landing, landing heavily again and overturning in one of the drainage ditches.

2/Lt. Edward Hotchkiss was accorded a burial with full military and civic honours, indicative of both an all-pervading patriotism that was yet to be eroded by the attrition of five years of war and the almost awesome reverence in which these pioneer aviators were held by the public at large. In the tiny village of Wolvercote flags were at half-mast as the coffins, borne by RFC colleagues, were carried from the Parish Hall to the church, escorted by contingents of the Queen's Own Hussars, the Royal Engineers, fifty members of the National Reserve and the band of the local Territorials. 'Top brass' from the Army Council, Admiralty and RFC led the mourners. For the short walk to Oxford Station the procession was brilliantly enhanced by additional escorts drawn from the Oxford Light Infantry, the Ox. & Bucks. Light Infantry, the Fire Brigade, the RAMC and St. John's Ambulance Corps. At the city boundaries the cortege was met by the Mayor and members of the Corporation in full civic regalia.

As the train passed slowly through Ludlow at 4.18 pm, dead on schedule, flags were lowered and the minute bell tolled. At 4.58 pm the body of Lt. Hotchkiss arrived at Craven Arms, where it was met by an escort, firing party and band of the Shropshire Regiment from Shrewsbury, and the Ludlow Company of the Territorials. A beflagged gun-carriage was provided by the Shropshire Royal Horse Artillery.

Interment was at Stokesay Parish Church, with which parish the Hotchkiss family had been associated for over 200 years as farmers and lime burners. A year later, on 6 December 1913, a stained glass window and brass was dedicated to the memory of Shropshire's 'first aviator'. The window, by James Powell & Sons, Whitechapel, London, is of two lights. The one depicts St. Michael armed with cuirass, sword, spear and shield on the Rock of Faith. In the other the Archangel Gabriel is represented as in the Hymn of the Nativity, with doves beneath his feet and the Holy Spirit above his head. The accompanying brass reads: 'To the Glory of God and in Memory of Lt. Edward Hotchkiss of the Army Flying Corps (the first Shropshire Aviator) who lost his life while flying at Wolvercote, Oxfordshire, 10 September 1912, aged 29. This window and brass were erected by his relatives and many friends'.

Six months earlier the *Whitchurch Herald*, as befitted Shropshire's leading aviation journal, had reported on the unveiling by Major Brooke-Popham of a granite memorial tablet set into the rampart of Wolvercote Bridge and paid for by over 2,000 subscribers. It was a joint civic/military ceremony 'town and gown' being represented by the Mayor and Corporation, the Master of University College and the Senior Proctor. Detachments of the Oxford TA, the Ox. & Bucks. Light Infantry, and No. 3 Squadron RFC formed a guard of honour. A Farman S.7 and a RAF BE2a of No. 2 Squadron, piloted by Capt. Dawes and Lt. Chinnery flew past during the two-minute silence. The tablet has a Bristol-Coanda monoplane incised in silhouette and bears the following inscription:

IN DEEP RESPECT FOR THE MEMORY OF
LIEUT. C. A. BETTINGTON & SECOND-LIEUT. E. HOTCHKISS
OF THE ROYAL FLYING CORPS WHO MET THEIR DEATHS
IN THE WRECK OF A MONOPLANE 100 YARDS NORTH OF
THIS SPOT
ON TUESDAY SEPT. 10 1912
SYMPATHISERS IN OXFORD AND WOLVERCOTE TO THE
NUMBER
OF 2226 HAVE ERECTED THIS STONE AS A TRIBUTE TO THE
BRAVERY OF THESE TWO BRITISH OFFICERS WHO LOST
THEIR
LIVES IN THE FULFILMENT OF THEIR DUTY

Two World Wars later there are literally hundreds of similar memorials to be found in churches and chapels everywhere, but none, perhaps, tell a story of greater significance to the development of military aviation in this country than that — if not the first, certainly one of the earliest such memorials — hidden deep in Shropshire's border country.

Fledgling wings

Military airfields, whether of First or Second World War vintage, were not sited haphazardly. More often than not they had an established, if sometimes tenuous, pedigree going back to the pioneering experimental years of civil aviation, even in areas remote from the main thrust of early flying. Thus, in March 1940, more knowledgeable locals could remark that at least three of 48 MU's dispersals at RAF Hawarden, Flintshire, had direct links with the earliest flying on Deeside — the 'Cop House Farm' site from which Cobham's barnstormers flew in the thirties; the 'Beeches', home to the Dutton family and birthplace of Thomas Murthwaite Dutton, aviation engineer and aircraft builder and reputed founder of RAF Sealand alias Shotwick; and the Sandycroft dispersal, located on the very fields used by RNAS and RFC planes as they sought the scarce facilities of Dutton's engineering works. Such coincidences, of course, do not completely explain the siting of RAF Hawarden — socio-economic and political factors, over-riding common sense, played the major role in the selection in 1937 of a large area of reclaimed marsh for an aerodrome — but they do serve to spotlight the lesser known men who, each in his own little patch, helped prod north east Wales into the aviation age.

Thomas Murthwaite Dutton was born at 'The Beeches' on 12 July 1881, second child and eldest son of Thomas Jenkins and Elizabeth Dutton, well-to-do farmers, employing five workmen and a governess to look after the growing family. Thomas Murthwaite was educated at Saltney, Hawarden and Chester, in which latter city the Dutton family anciently had its roots. In the light of Thomas Dutton's later association with Caudron aeroplanes one can draw interesting parallels with the French Caudron brothers, René and Gaston. All three were born within a couple of years of each other — Gaston Caudron on 18 January 1882, René Caudron on 1 July 1884, at Favières, north of the Somme estuary, also of farming stock. All three developed an early interest in things mechanical.

Theirs was an age where the peace of the countryside, so long broken only by the clip-clop of horses' hooves and the resonant roll of iron-tyred wheels, or the staccato rumble and whistle of trains, now established over half a century, was increasingly disrupted by dusty clouds marking the passage of the new-fangled automobile along brown and gritty roads. Rich young men — and the not so young — suitably attired in hairy coats and goggles, found a new freedom and status on wheels, surpassing that accorded to the masses by the £3 bicycle! The twenty-year old Thomas Dutton was familiar with Hal Hulbert's (Llwyn Offa, Mold) Pick, Frank Summers's (Bromfield Hall, Mold) Napier and Gobron Brille, E. S. Taylor's (Little Acton, Hawarden) Gladiator, and George Armstrong Parry's (Oaks Farm, Buckley) Humberette.

But because of the 20 m.p.h. speed limit in England real thrills were confined to specialised tracks such as Brooklands (built 1906/7 by the Hon. Hugh Locke King at his own expense) or abroad, particularly in Ireland. Flintshire's own particular knight of the new age was Vivian Hewitt, ultimately heir to the Hewitt Bros. Tower Brewery (Grimsby) millions. Although educated at Harrow he was not permitted to live a life of idleness. No financial support was forthcoming until he had proved himself capable of earning his own living. Like many of his contemporaries Hewitt turned enthusiastically to his hobby, engineering. A stint at the Marine Engineering Department, Portsmouth Dockyards in 1904 was followed by four years as a 'privileged apprentice' at the LNWR works at Crewe, mastering track laying and maintenance, the intricacies of signalling, as well as every aspect of the steam locomotive. He frequently served as supernumerary fireman on the footplate of Crewe-Holyhead expresses, a welcome change from the engine room of the LNW Irish Mail ships out of Holyhead. The ownership of two motor-bikes — a Triumph and an American Henderson — and his appointment in 1910 as an official driver for Singer racing cars further fuelled his passion for engines and zest for speed, given tangible utterance by successful forays into the second-hand car market, buying and up-grading high-performance machines and selling them at large profit to like minded devotees of the 'Goddess of Speed'.

But with Blériot's crossing of the Channel in 1909 the starry-eyed engineer/would be entrepreneur was lured into the even more adventurous world of the aviator. Through his uncle's benevolence

A young Vivian Hewitt with his home-built glider at Bodfari.
[Raymond Davies]

Hewitt's Brooklands business card, 1909.

Hewitt became the proud owner of an 'Antoinette' high-winged monoplane, powered by a 50 h.p. in-line engine, capable of speeds up to 44 m.p.h., and costing £1,000. But while prominent at every aviation meeting between 1910–11 the dragonfly-like machine was not a good flyer even in the most perfect weather conditions. Small wonder, perhaps, that in 1912 the company went into liquidation. In 1909 Hewitt had rented a flying-shed at Brooklands where many aviation pioneers such as A. V. Roe, Short Bros. , and Martin and Handasyde had taken up residence. The aerodrome was in the middle of the racing track. Here Hewitt had the best of both worlds. In 1912 Vickers opened their flying school at Brooklands, which became the focus of the aviation world before the outbreak of the First World War — despite the proximity of high tension cables, factory chimneys, wooded hills, built-up areas and railways!

In January 1910, following another appeal for family funding, Hewitt swapped his 'Antoinette' for a Blériot monoplane with a 25 h.p. rotary le Rhone engine, costing £1,100. In February 1911 this would be upgraded to the larger Blériot XI with a 50 h.p. Gnome engine with which he would thrill the holiday crowds of the Flintshire and Denbighshire resorts. Strangely, although Blériot also had a flying shed at Brooklands, Hewitt had to travel to Pau in the foothills of the Pyrenees, where Blériot Aeronautique had its works and aviation school, to pick up his monoplane. His business cards and company notepaper would henceforth read 'Dealer in Second Hand Cars and Aeroplanes… Repairs to Blériot Monoplanes a Speciality. Wings Constructed, etc. '

Before perhaps making a fool of himself at Brooklands, trials in his new monoplane were carried out at Bodfari. Why? one will never know, because he had already concluded in his glider building days that the narrow steep-sided Wheeler valley with its unpredictable air currents, as not exactly the best flying terrain. And so it turned out. On its second flight the Blériot's engine failed at 300 feet, and crashed alongside the LNW railway line just outside Bodfari station. Pilot was severely concussed and the Blériot's frame badly smashed. Thomas Murthwaite Dutton was called in, the first of several visits to Warren House, to assist in repairs. Within a month both machine and battered owner were back at Brooklands.

At Brooklands, since aviation was in its infancy and the object of great public interest, aviators, Hewitt amongst them, would take to the air on car racing days, adding to the thrills and spectacle. This turned out to be quite lucrative. Therefore in June 1911, having taken his pilot's 'A' licence, Hewitt forsook Brooklands for Flintshire where flying was virtually unknown and the potential for profit from exhibition flying seemingly limitless. Again exhibiting an independent streak, he moved out of Warren House, took digs in Rhyl, and established a landing strip at Foryd Fawr on the west bank of the Clwyd. Christened 'Foryd Aerodrome' this field would become the

Vivian Hewitt, 'Rhyl's First Aviator', stands in front of his immediately identifiable Blériot monoplane on the beach at Rhyl, surrounded by spectators. [FRO]

PUBLIC WELCOME
TO
VIVIAN HEWITT
ESQ., AT THE
RAILWAY STATION, RHYL,
TO-NIGHT
(Monday, April 29th)
ON HIS
Arrival from Holyhead by the 10 o'clock train.
Public Reception
afterwards in the
NEW PAVILION

Chairman   -   Councillor J. E. BUCKLEY JONES, J.P.
(Chairman of the Council).

Rhyl Town Band
and TORCHLIGHT PROCESSION.
The Record and Advertiser Co., Ltd., Rhyl.

Above: Vivian Hewitt in the cockpit of his Blériot XI. [FRO]

Left: Poster for the ad hoc *welcome to Vivian Hewitt just three days after his successful crossing of the Irish Sea. [FRO]*

Public Presentation to Vivian Hewitt, Esq.
In recognition of his Flight
From Rhyl to Phœnix Park, Dublin, on April 26, 1912
(A World's Record for Over Sea Flying).

Thursday, August 1st, at 4-30 p.m.,
IN THE
NEW MARINE GARDENS.

Kindly present this Invitation at the Entrance Gates facing Pavilion, as a space on the Terrace will be reserved for invited guests.

R. S. C. SYKES, Chairman and Hon. Treas.
J. D. POLKINGHORNE, Hon. Sec.

Above: Invitation to the Public Presentation to Vivian Hewitt on 1 August 1912.

Left: Presentation of an illuminated address to Vivian Hewitt at the Rhyl Pavilion. [Rhyl Library]

base for his seasonal exhibition flights over the North Wales and Wirral resorts. A large shed was was converted into what *The Aeroplane* described as one of the best equipped workshops in the country. It was in this shed that the paths of Hewitt and Thomas Murthwaite Dutton would cross again.

Hewitt's first flight over Rhyl took place on Thursday, 12 October 1911. This was prudently cut short owing to engine trouble. But on Wednesday, 18 October, he made two 'splendid flights' over the town, passing over the Marine Lake and St. Thomas's Church and out to Towyn-uchaf before flying back along the sea front to the acclaim of large crowds. Thus began the profitable association with the town of Rhyl which would last until the outbreak of war in 1914. 'Brilliant flights', 'beautiful flights', all made to the apparent benefit of Rhyl's ratepayers 'and to the envy of other towns on the coast' (especially Abergele!). By April 1912 his new Blériot XI, immediately identifiable by the words 'VIVIAN HEWITT' painted on the underside of the wings, had clocked up some 6,000 miles, the greater part in attending minor flying meetings in Lancashire and Cheshire to which he had managed to wangle an invitation. He had yet to make a name for himself. Locally flights were usually made along the coast and the Dee estuary, rarely inland up the Vale of Clwyd. Indeed, his first flight round Rhuddlan Castle and on to St. Asaph, where he circled the Cathedral (and disrupted a Parish Council meeting), was only made on Tuesday evening, 16 July 1912. Apparently treacherous air currents in the Vale had to be treated with respect and caution, but possibly the real reason was that there were no Chester-bound express trains which he could race!

Vivian Hewitt's finest achievement was the first successful crossing of the Irish Sea on 26 April 1912, at some seventy-five miles a flight three times further than Blériot's Channel crossing. Following Robert Loraine's near miss on 11 September 1910 when the unfortunate aviator's Farman came down in the sea off Howth Head at the mouth of the Liffey, Hewitt had always harboured this secret ambition. His hand was forced in April 1912 when aviators, both experienced and inexperienced, seemed to be queuing to attempt the crossing. When, at 06.00 hrs. on Thursday, 18 April 1912 Leslie Allen took off in his Blériot from the Roodee at Chester to follow the LNW railway line to Bangor and Holyhead and thence out to sea, Hewitt and Sydney Wingfield, his mechanic, were stranded in London buying spares for their own machine. They could only wait on reports whether or not Allen had been successful in his crossing attempt. Amongst the interested spectators who logged Allen's optimistic passage through the early morning mists shrouding the salt marshes along the Flintshire bank of the Dee estuary was one Thomas Murthwaite Dutton.

Banner headlines in the evening newspapers 'Airman lost in Irish Sea' proclaimed Allen's fate. Hewitt and his mechanic raced back overnight to Foryd where his Blériot lay dismantled, convinced that another crossing attempt was imminent. Calling again on the assistance of T. M. Dutton as a third mechanic, they worked feverishly to re-assemble the machine. At 02.00 hrs. on the Sunday work was complete. Early in the morning he flew from Rhyl via Llannerchymedd to the same landing field at Penrhos Park near Holyhead used by Robert Loraine two years previously. But here he had to wait until strong winds and sea mist had abated. At 10.30 hrs. on Friday, 26 April, Hewitt finally took off from Holyhead, watched by a large crowd. Without a compass, navigating occasionally by the sun as it broke through the thick haze, he finally made landfall near Bray Head off the Wicklow Mountains, some fifteen miles off course. It was a simple matter to follow the Wicklow coast north to Dublin, where at 11.15 hrs. he landed near the Wellington Monument in Phoenix Park, the first airman to cross the Irish Sea.

On Monday, 29 April 1912, following the return of the successful team from Dublin there had been a spontaneous public welcome and reception by the Rhyl Ratepayers' and Rhyl Advertising Associations in the New Pavilion. In the New Marine Gardens on 1 August, at the height of the holiday season, Hewitt was presented with an illuminated address on behalf of councillors, magistrates, residents and visitors 'In recognition and admiration of your courage and skill in flying the Irish Sea from Rhyl to Phoenix Park, Dublin via Holyhead, a distance of 75 miles (thereby establishing a World's Record in Aviation over Water) on the 26 April 1912.'

Rhyl basked in the reflected glory. 'Although Rhyl was but a small town in North Wales, Mr. Hewitt's great achievement has brought it to the notice not only of the people of Great Britain but to the whole world as being associated with a young but experienced and daring aviator'. Many endorsed the remarks of the Chairman of the UDC. Some were more visionary than others. 'We rejoice in his success and we hope and believe that when the future uses and benefits of aerial locomotion are realised, the name of Mr. Hewitt and Rhyl will still loom largely as its pioneers (loud cheers).'

Brave prophetic utterances, doomed to unfulfilment. The name Vivian Hewitt is lost to the standard pages of aviation history, while Rhyl's fragile lead in 'aerial locomotion', real or imagined, would be eroded by the petty rivalries, jealousies, squabbles, and intrigue which were the hallmark of local government along the holiday coast. Indeed, one might ruefully reflect that thirty years later plans to establish RNAS aerodromes at Abergele (torpedo, bomber, reconnaissance) and Rhyl (fighter) were fiercely resisted by councils and a vociferous farming lobby. Again, in June 1962, with the inauguration of an experimental hovercraft service between Rhyl and Wallasey, the *Rhyl Journal* could comment that for the second time Rhyl was occupying 'a position of supreme importance in the history of travel'. Comparisons were made with Hewitt's historic flight some fifty years earlier; indeed the Vickers V.A.3 hovercraft was based in Foryd harbour barely 200 yards from Hewitt's former aerodrome. But once more optimism was misplaced and the service was withdrawn six months later.

Hewitt's successful 'publicity flights' over Rhyl, in which he dropped specially signed and dated cards over the throngs of holiday-makers, had begun to lead to flying engagements elsewhere in Wales, notably rural areas as far south as Radnorshire, well outside Gustav Hamel's itinerary. But following the Irish Sea crossing he became overnight a 'national' as distinct from 'local', aviation figure. Invitations to give flying exhibitions came flooding in, but although he could now afford to pick and choose, Welsh agricultural shows remained a favourite venue! Indeed, he had to appoint an agent to oversee his engagements in South Wales!

With the outbreak of war in 1914 Hewitt's tiny aerodrome at Foryd was closed and he was commissioned as a Acting-Lieutenant into the RNVR. His last appearance on the flying circuit was possibly that at the Rock House Hotel, Llandindrod Wells, where, on Wednesday, 5 August 1914, he met with disaster as his Blériot caught the top of some trees on a wooded hill and crashed to the ground, smashing the propeller. That was that, a suitable point to call it a day. ! The same day a local printer rushed out fly-sheets announcing that 'as Mr. Vivian Hewitt (who is giving an Exhibition of Flying today in the Rock Park Grounds) has Volunteered his Services to the War Office, he will be UNABLE TO FULFIL HIS SECOND ENGAGEMENT to fly at Llandindrod Wells in September'. The next twelve months was spent testing aircraft at Farnborough. In November 1915 Hewitt went to the USA with the British War Mission, flight testing aircraft for the Admiralty at the Curtiss Aeroplane and Motors Corporation flying fields at Buffalo and Hammondsport, New York and Newport News, Virginia. Unfortunately a bad crash in January 1918 would mark the end of his flying career, and he was retired from the RNAS with the rank of Captain.

Although parallels may be drawn between the two, Thomas Murthwaite Dutton's career was very much a lower key affair.

However, he would have a greater residual impact on the local landscape. As a budding mechanical genius he was apprenticed to Taylor's Engineering Works, Sandycroft (later Sandycroft Foundry and Engineering Works), which firm manufactured steam and electrical plant to customers' specifications world wide. The firm had a long industrial pedigree, the foundry being originally established by John Taylor, the mining engineer, at Rhydymwyn in 1837 in connection with the Mold lead mines. It had moved to Sandycroft in 1862, continuing to make mining machinery and, from the 1890s, electrical motors. It would close in 1925 during the post-war economic slump.

Three years ahead of Dutton, one Alfred Ernest Owen (b. 1869), of 'Woodley', Sontley Road, Wrexham, had also completed his apprenticeship at Sandycroft. His father, Alfred Owen, ran an agricultural implement warehouse and ironmongers in Chester Street, Wrexham, and no doubt the premium apprenticeship at Sandycroft was secured with the long-term interest of the firm in mind. But the young man had different ideas and struck out on his own. In 1891 John Turner Rubery had established a small steel constructional company at Darlaston. Consolidation and expansion was possible only with some financial and technical input by a partner. In July 1893 A. E. Owen, with a modest loan from his father, answered the advertisement. In 1905 the name of the firm became Rubery Owen & Co. Five years later J. T. Rubery retired, selling his interest to Owen.

Already doing pioneer work for the automobile industry, it was but a sideways move to meet the demand for specialised materials — high stressed nuts and bolts, engine attachment plates, bracing plates and wire etc. — from the rapidly expanding aviation industry. In 1909 the firm is noted as supplying propeller plates, steel, bolts and wire for 'the new all-Welsh aeroplane being constructed [at Llanddona, Anglesey] on exact and profound mathematical principles' by Ellis Williams, an assistant lecturer in physics at University College, Bangor. Thirty years later the twenty-eight companies in the Rubery Owen Organisation were producing components for almost every type of aircraft currently in use by the RAF, whilst also erecting and running the MAP propeller stamping factory at Whitegate, Wrexham in which they would relocate from Darlaston in 1946.

Thomas Murthwaite Dutton at the wheel of Model 'T' Ford 'DMT4' (a customised number plate produced solely for purposes of this publicity shot) in front of his garage at Queensferry. His aero-engine workshop was to the rear, [Simon Warburton]

E. A. Owen had a flair for improving on existing products — making them more reliable — and the quality of materials as well as up-dating and refining production techniques. Patent Office records show Owen, along with colleagues at the Darlaston works, registering some forty patents between 1905–1940, such as 'Improvements in Sheet Metal compression members for Aircraft such as are used for Interplane Struts [wings], Chassis Struts [undercarriage] and Wheels' (3 May 1923) and 'An Improved Method of constructing the Central Portion of Wire Wheels for all Vehicles [including aircraft] comprising the Hub and Spoke Flange' (July 1929).

In the meantime, numbered amongst Rubery Owen's customers, was Thomas Murthwaite Dutton, beginning to make his way in the aviation world and now living at 'Rosleigh', Vicarage Road, Hoole, Chester. Until June 1913 Dutton worked at the Sandycroft foundry as an electrical engineer, becoming an AMIEE in 1910, but his growing interest in aviation prompted a bold change in direction. He regularly attended flying meets in Cheshire and Lancashire although as yet a non-flyer (or at least an unqualified one!). As he became known and accepted on the flying circuit his electrical and mechanical expertise was often called upon in emergencies. The key to success at these early 'fly-ins' was not so much a good airframe design as engines which started easily and ran reliably and Dutton became known as one able to undertake any sort of engine repair at short notice. Such contacts served him well. Through the generosity of friends he was able to log up many hours of 'air experience' and quickly became quite an accomplished pilot hacking about in an assortment of aircraft. This sowed the seeds of an ambition, as money was forthcoming, (a) to establish his own aircraft construction business and (b) like so many of his contemporaries, to run his own flying school in conjunction with it, the one to finance the other. Three years later, under the impetus of war he would achieve both these aims and become a qualified pilot into the bargain. But first things first.

In 1913 he was encouraged to set up his own engineering workshop in Queensferry, repairing and rebuilding aero engines, and moved to Upper Dale House, Hawarden. For the moment this by itself was insufficient to keep him and by 1914 his aero engine workshop was fronted by a fully fledged automobile service garage — T. M. Dutton, Queensferry Garage & Engineering Works — sole agents for Cadillac, Overland Ford and Hubmobile cars, sited strategically at the junction of the A548 and B5129. He also had a subsidiary workshop in Park Street, Chester, where repairs were made to car and motorcycle engines.

In making this move he was following, perhaps unwittingly, the steps of more exalted aviation pioneers, both at home and abroad. Blériot was an engineer and mathematician and came to aviation from automobiles. Louis Bréguet, one of the first designers to produce a satisfactory tractor biplane, was an electrical engineer before forming his own aeroplane company, the Societé Anonyme des Ateliers d'Aviation. Henri and Maurice Farman, inventors and test pilots, had a background in bicycle building, car sales and racing before establishing their huge integrated aeroplane factory at Billancourt (Seine) in 1912, two miles long, employing over 5,000 and turning out ten aircraft a day. The oldest aeroplane firm in the world, Aeroplanes Voisin, was founded in 1905 by Charles and Gabriel Voisin, inventors and builders of anything electrical and mechanical. These were the 'big frogs' in an expanding pond. Dutton would remain perhaps 'a little frog' in the aviation puddle that was north-east Wales. However, the coming of war would see him grow in stature and his 'puddle' into a considerable sized 'lake'.

In 1912, whilst on the aviation circuit Dutton met up with William Hugh Ewen, a Scot, since 1910 proprietor of a one-plane 'flying school' at Lanark Racecourse but also, as W. H. Ewen Aviation Ltd., holder of the sole UK rights to build and supply Caudron aircraft. René Caudron had his factory at Issy-les-Moulineaux (Seine) and a flying

Dutton's aircraft shed, Queensferry (South Shotwick) c.1916. A surplus Avro 521 undergoes overhaul with two Caudron G.3s in the background. [FRO]

school at Le Crotoy (Aerodrome de la Baie de la Somme). His aircraft were some of the most widely used French aeroplanes in the First World War, especially his twin engined bombers noted for the weight lifting capacity and imperviousness to bad weather. In 1913 Ewen was supplying Caudron amphibian seaplanes to the RNAS, expanding in the first year of the war to include Caudron G.2s, G.3s and G.4s. By January 1913, Ewen had moved his school and aeroplane assembly operations to a shed on the rapidly developing 'London Aerodrome' at Hendon. Here Dutton worked on, and flew, two early Caudron G.3s, Nos. 40 and 45, as yet only with 80 h.p. Gnome engines, which might explain his preference for this aircraft when he established his own little assembly plant back in Flintshire.

With the outbreak of war in August 1914 Hendon was requisitioned by the RNAS as an Aircraft Park, receiving and distributing new aircraft from the manufacturers. Ewen's flying school, along with four others on site, were contracted to train pilots for the Services but would continue to take civilian students alongside the military until September 1916 when the flying schools were absorbed into the RFC's School of Instruction. Dutton must have been one the last civilians to pass through the School.

The Ewen-Dutton-Hendon link might go some way to explain the sudden appearance in Queensferry in early 1915 of a RNAS detachment who set up a canvas Bessoneau hangar in a field south of the Sandycroft road near to Dutton's workshop, a field later to be used as a RLG for Shotwick and in the Second World War as one of 48 MU's (Hawarden) dispersals. Older residents recall both a RNAS and RFC presence, with bi-planes landing from time to time and crated aircraft both arriving and departing on motor lorries and horse-drawn wagons. Aero engines were dismantled and taken to Dutton's workshops for testing and overhaul before being shipped out, and one can only assume that this aero-engineering facility was the prime locating factor in the establishment at Sandycroft of what amounts to a

small temporary storage/packing depot. Official records do not shed light on the depot's function at that moment in time. The First World War aerodromes at Shotwick, Hooton Park and further afield at Tern Hill and Shawbury would not come into being until the latter half of 1917. As will be seen in the following chapter, the only operational station aerodrome in North Wales was the RNAS Llangefni, which would come on stream in September 1915. But this was essentially a Naval Airship Patrol Station, complete with its own workshops. The Sandycroft depot/aerodrome was can only have been connected with the shipment of aircraft from Birkenhead docks to distance theatres of war *e.g.* to the East African Field Force and the Mafia and Niororo Islands, to which shipments of Caudron G.3s had started as early as July 1915.

With these Government contracts and possibly more in the offing, in 1915 Dutton leased some 142 acres of former marshland from John Summers & Sons Ltd, the adjacent steelworks company which had purchased over 10,000 acres of estuarine saltings, much of it yet to be reclaimed, to secure any future development. It was also a period of expansion on the automobile front. Dutton took new premises at 52 Bridge Street, Chester and adding Ford vans to his dealership. This side of the business was put in charge of a Mr. G. Williams. His advertisements now read 'T. M. Dutton, AIMEE, Queensferry and Chester'.

His piece of land at Sealand was the whole of that area reclaimed in 1833, the seaward embankment being used in 1890 to carry the Great Central Railway from Hawarden Junction into Chester Liverpool Road. Drainage and cultivation made for compaction and shrinkage so that its surface was only 11–17ft above mean sea level. At one time, before further (post–1892) reclamation, the 'polder' was actually some 6–12ft below the level of the outlying salt marsh. Additionally the area was traversed by several shallow depressions marking the course of former tidal creeks. A lot of work, therefore, had to be carried out to

North Shotwick (Sealand) from the north-west with (centre) the single 'Belfast' aeroplane repair shed that was Dutton's domain 1917–18. [FRO]

Caudron G.3s ready for flight testing at South Shotwick. The hedge line marks a dyke of 1833. It is straddled by Dutton's aeroplane shed which thus had access to the fields on either side. [FRO]

Below: RAF Shotwick from the south-east, c1920. In the foreground (south of the railway) 'Duttons Flying Ground' emerges as an Air Acceptance Park (South Shotwick). The other side of the railway is North Shotwick flying ground.

make the area suitable for flying. A central dyke, running roughly east-west and lined by a 12ft hawthorn hedge or wind-breaker, divided the proposed aerodrome into a North Field and South Field. Straddling this dyke lengthways on piles Dutton built an enormous aeroplane shed, 350ft long, 31ft wide (discounting overhanging eaves) and of modest height, 11ft clearance to the roof trusses, giving uncluttered floors with adequate lighting through central lighting in the roof apex. Learning from the best of current practices as observed at and about Hendon, Dutton was already thinking in terms of a simple flow-line production system, even for the comparatively small numbers of aircraft he would build and handle. A section of the shed was fire-walled and ventilated for storing dope, paints etc. There was very little in the way of machinery as most of the structural and mechanical engineering was done in in Dutton's workshops in Queensferry. Access to the construction shed was by ramps at each corner of the building.

By August 1916 Dutton's Sealand flying field was in use mainly by friends and clients, he himself flying in a modified Caudron G.3 with its second-hand reconditioned 80 h.p. Le Rhone engine, the first aircraft turned out at his new 'factory'. At the end of September he travelled to Hendon and enrolled with the J. Laurence Hall School of Flying, now busy churning out RFC pilots. W. H. Ewen had earlier obtained a commission in the RFC. At Hendon his instructor was Donald Clappen (later Air Commodore), a pre-War pilot and formerly an instructor with the Blériot Flying School when it was based at Hendon. It was therefore no coincidence that in 1917 Major Clappen, now a field officer for the Directorate of Aircraft Depots and Acceptance Parks, should descend upon Sealand in his search for AAP sites within easy striking distance of the Merseyside ports.

Dutton began his training on a Caudron G.3, moving quickly on to Hall's impressed Avro 500, the Avro 504 and the Airco DH6. On 24 October 1916 he duly gained his RAC Aviator's Certificate (N° 3,730). The examination consisted of flying five figure-of-eights at varying distances from the airfield, an altitude flight, a stall, and a landing without power. During the six weeks he was at Hendon Dutton spent some time at the British Caudron Company's sheds (Ewen's successor firm), as well as visiting other manufacturers in and around Hendon — Grahame-White, Handley Page and the Aircraft Manufacturing Co. — picking up tips on organisational and production methods that would stand him in good stead on his return to Flintshire.

He was particularly interested in the development of the compressed air catapult that would revolutionise the operation of aircraft from ships at sea. Between June November 1915 F/Lt. R. E. Penny had been prominent in trials flying from the deck of a ship. In October 1917 F/Cdr Penny, in a specially strengthened Avro 504C (upgraded to 504H), would become the first pilot to be successfully launched from a compressed air catapult. The latter was designed by R. F. Carey and built by Waygood & Otis. But it is interesting to note, and older residents of Deeside recall quite clearly, that Dutton tried to muscle in on the idea, and in conjunction with Sandycroft Engineering, his former employers, embarked on a research project to produced a catapult propelled by an electric motor. A prototype was erected on the banks of the Dee and tested with a mock up 'box' or crude aeroplane fuselage filled with scrap and propelled across the New Cut on to the opposite bank, much to the chagrin of small holders and market gardeners! These experiments went on for some time until abandoned in the light of (a) Penny's Hendon success and (b) the threat of legal proceedings arising from the obstructing the navigable channel of the Dee!

Dutton's return to Queensferry coincided with the first storms of winter and subsequent flooding which made his new aerodrome unfit for flying. Fortunately there was the field in Sandycroft near his garage. The latter would also serve as a RLG to 5 FTS in the 1920s and early 1930s when RAF Sealand was under water! However, Dutton made good use of his time by perfecting the design of his modified

Caudron G.3s. Most of the parts were made in his engineering workshop. Wood was obtained from Chester (J. Musgrave, St. John's Street) or Connah's Quay (W. M. Butter, T. J. Reney) timber importers; airframe fabric from United Mills Ltd., Northgate, Chester, sail-makers and tarpaulin manufacturers, while A. E. Owen supplied strut bracing irons, rigging wire etc. His second-hand Le Rhone engines were obtained from Government surplus stores or from friends who had enlisted and laid up their aeroplanes. These engines Dutton rebuilt at Queensferry.

In February 1917 Dutton was elected a member of the Royal Aero Club. His aerodrome was drying out and flying recommenced. He now advertised himself as 'Murray (his nickname!) Dutton's Flying School, Sealand, Queensferry. All aspects of Flying Tuition undertaken. Service personnel trained. ' In aviation journals he also appears, more up market, as 'Dutton Aircrafts [sic], 52 Bridge Street, Chester'. Interestingly, in early 1919 he was advertising as 'T. M. Dutton AMIEE, Engineer, Designer and manufacturer of Aeroplanes. Head Office: The Aerodrome, Sealand, near Chester'. Dutton, reflecting popular local usage, was therefore the first to give the name 'Sealand' to the establishment that until 1924 was known officially as 'RAF Shotwick'.

The building of Caudron G.3 types was possibly due not only to Dutton's early experience with this type, but also to the fact that René Caudron, as a French patriot, had waived licence fees for other manufacturers to help prosecute the war in the air and to provide aircraft for training. His aircraft building operations for the moment provided planes for the still flourishing civilian market, but his Flying School was now geared to meeting military needs, although not taken over. Private flying was largely confined to the field the other side of the river in Sandycroft. In March 1917 his aerodrome had been inspected and approved by RFC officers, another possible reason why it was on file as a possible AAP. His fees were 50 guineas for a complete flying course in a bi-plane. His students were either serving officers seconded or volunteered from the Cheshire Regiment or Royal Welsh Fusiliers, or civilians who wished to enlist in the RFC but who first had to learn to fly *ab initio* — and, if accepted, would have their tuition fees refunded! Such were the difficult first steps on the road to glory — and an average life expectancy of two weeks with an operational squadron on the Western Front!

The two sons of John Percival Gamon, Chester solicitor, secretary of Chester Golf Club at Sealand, clerk to Chester City Council and post-war a sleeping partner in Dutton's car business, typify the intake at the latter's flying school. Sydney Percival Gamon held a commission in the 1/5 Battalion (TA), the Cheshire Regiment, commanding the machine gun section. His application in August 1914 to join the RNAS was unsuccessful. He was with his regiment in France between February 1915-May 1916. Now substantive Captain, he transferred to the RFC and acted as observer. In November 1916 he returned to England to learn to fly, completing his initial training at Dutton's flying 'academy', thence to the Reserve Aeroplane Squadron at Farnborough before being posted to 78 (Home Defence) Squadron, Sutton's Farm (Hornchurch post–1928). From here, on 29/30 January 1918 he flew his first anti-Gotha patrol. Unfortunately on 28 March 1918, whilst flying a Sopwith F1 Camel (C6726) he lost control during aerobatics, spun in and was killed. He was buried with full military honours at Neston-cum-Parkgate Cemetery, the guard of honour being provided by men from his old regiment and the RFC at Shotwick.

His younger brother, John Gamon, was commissioned into the RNAS as a PBSL on 30 July 1916 and flew at Sealand before advanced training at Chingford, Essex, and Cranwell. In July 1917 he was flying Sopwith F1 Camels with 4 Squadron at Bray Dunes (French frontier). On 30 March 1918, now a F/Lt. with 5 (later 205) Squadron flying DH4 bombers out of a quick succession of shelled out airfields, he fought the action which was to gain him the DSO. As the *London*

Gazette of 7 June 1918 records: '… for conspicuous gallantry and devotion to duty. On 30 March 1918 whilst returning from a bombing raid he was attacked by three enemy tri-planes, (Fokker Dr.Is) one of which he brought down and drove off the other two. He has also carried out many bombing raids on enemy lines of communication, aerodromes and ammunition dumps. His work has always been of the greatest merit and he has set a splendid example to those around him'. On this occasion Gamon was flying DH4 serial N6004 with OSL F. H. Stringer in the rear cockpit. On 24 April 1918, with F/Lt. R. Scott on aft Lewis gun, he would bring down another Fokker Dr.I in flames and send a Pfalz D III falling out of control at Chaulnes railway station. On 17 June 1918 Capt. Gamon was injured when his brand new DH4 (D9277) was hit by flak whilst on patrol.

From October 1917, with the arrival of the first RFC units at North Shotwick, Dutton found himself contracted to repair RFC/RAF machines. There are extant for this period several interesting photographs of the inside and outside of his aircraft shed and the appearance of RFC roundels on wings has always intrigued local historians. However, one of the airframes, minus propeller and roundel embossed wings (the latter propped against a wall), is in the process of having new fabric applied, and from the 110 h.p. Clerget engine with its distinctive cowling has been identified as an Avro 521, a rare 'bird' indeed. This was a two-seater fighter single bay variant of the Avro 504. Test flown and evaluated in February 1916 a production order for twenty-five followed (serials 7520-7544), but although a number were built the type proved unsatisfactory and none entered service. Built at A. V. Roe, Miles Platting, Manchester, it would seems as if at least one redundant aircraft reached Dutton's works at Sealand, possibly for some modification before selling on.

Business may have been such that the Great Central Railway was moved to provide a halt and a short single siding on the northern perimeter of the flying field, very much in the same way as it had supplied at least three platforms for the adjacent Chester Golf Club. Railway records do not have a precise date for for Welsh Road Halt (known as Sealand only from 14 September 1931) except an enigmatic 'post 1915'. If 1916 the halt belongs to the Dutton era. But since the same records confirm 'restricted (*i.e.* military only) use' to '*c*.1919', the halt may have been built to accommodate first contractors' men and then RFC personnel at North and South Shotwick. From a non-railway source (evidence given at a coroner's inquest following a fatal accident at the station) the opening of the halt to the general public may be dated more precisely to Monday, 12 August 1918. On 31 August one Margery H. Tucker, having just alighted from a passenger train, had been killed whilst crossing the line by an engine shunting out the siding. Daughter of the vicar of Stoke Courcy, Somerset, she had enlisted in the WAAC (WRAF since 1 April) and worked in the paint and dope shop on North Shotwick. WRAF quarters were south of the railway, hence the tragic accident. She was buried with full military honours in Chester Cemetery.

Dutton clung on to his rather anomalous position on his flying ground south of the railway, but his independent days were numbered as giant Belfast hangars, messes and barrack huts took shape on the eastern side of the field opposite Old Marsh Farm. A putative Air Acceptance Park to deal with an anticipated influx of American aircraft through Birkenhead, it never received an official number (for that matter, not many AAPs did!). The end of the war interfered with full commissioning, but, as will be seen later, it did have some limited use. On 16 October 1918 Dutton was commissioned as Captain into the RAF, belatedly to regularise his position as being in charge of engineering and repair at 37 Wing's peculiarly (but aptly) named 'Half ARS' (Aeroplane Repair Section). The other 'Half-ARS' was at Hooton Park.

Published details relating to 37 (Training) Wing RFC are often conflicting, but it was formed on 15 October 1917 ostensibly to have oversight of 4 TDS (Hooton Park) and training units at North Shotwick. HQ was initially at 'The Oaks', Ledsham, near Little Sutton, roughly half way between the two aerodromes, before moving into a purpose-built HQ buildings, firstly on South Shotwick, and then, from 3 November 1918 at Hooton Park. It would disband on 9 April 1919 six weeks before Hooton Park closed. Its functions were taken over by 13 (Training) Group upon which would evolve the responsibility for running down and closing most of the RAF's installations in the north-west. Significantly Capt. T. M. Dutton went on to the Reserve List on 8 April 1919.

Dutton remained at Sealand hoping to recover some, if not all, his aerodrome as and when the station was derequisitioned. But it was not to be. Some flying took place as advertised from the South Field, but the AAP was busy handling not only running repairs for 4 TDS and 51 TDS but also storing and processing a great number of redundant aircraft., so many in fact that the now deserted Hooton Park aerodrome was also called into use as a sort of sub-sub-storage depot.

It is interesting to reflect that on 14 January 1919 Chester City Council approached the Air Ministry with view 'to getting their aerodrome back', that is, seeking a date for the reversion of Shotwick (Sealand) to civil use. This is the first piece of evidence to indicate that Dutton's airfield had also been *de facto* the aerodrome for Chester. On 14 March 1919 the city fathers were informed that Shotwick was 'a permanent RAF station' and that any use by civil aircraft would be in emergencies only. Thus a firm decision in principle on the retention of both sites at Shotwick by the military appears to have been taken very early. More detailed discussions on the logistics and strategies involved did not get under way until September 1919 and were not concluded until October 1920, although No. 5 FTS had been formed in April 1920 and would remain at Sealand until November 1940. The rail strike of September/October 1919, in which Shotwick AAP played a crucial role as an emergency distribution and collecting centre for mail, no doubt helped mould opinions. In the face of so many factors stacked against him Dutton decided to call it a day and concentrate on his automobile business, diversifying into motor-cycles and farm tractors etc.

So ended, rather lamely perhaps, the aviation career of one of Flintshire's unsung pioneers of the air. Amateur enthusiast, mechanical and electrical genius, self taught pilot, aircraft builder, flying school proprietor and chronically under-funded and little appreciated — these were exactly the same attributes possessed by others who went on to become household names — Alliot V. Roe, Frederick Handley Page, Geoffrey de Havilland, 'Tommy' Sopwith. Had not Queensferry been 'out in the sticks', had he not lost his aerodrome, who knows to what heights Dutton may have soared in the aviation world?

As a postscript one should perhaps note that in the face of the irreversible decision by the Air Ministry, Chester City Council decided to approach the Duke of Westminster's agent with the hope of being able to lease or purchase a piece of land for a new aerodrome nearer the city centre, but agreement was not forthcoming. On 3 June 1919 the city surveyor was asked to examine a site at Cop House Farm, Saltney — scene of city-sponsored barnstorming displays in the 1930s and of a Second World War dispersal unit for RAF Hawarden — but the Hawarden Castle estate would not release the land. Here the matter of a civil aerodrome for the city of Chester was left in abeyance until further (equally abortive) efforts were made between 1932–39. It may be that attention was drawn to the Cop House site by Thomas Murthwaite Dutton. After all, it was next door to Beach House, the Dutton family home. If Chester City had started up an aerodrome there it was on the cards that Dutton somehow would have been involved in the running of it!

The Llangefni 'Pigs'

Shortly after 15.00 hrs. on Friday, 26 April 1918, holiday makers on Llandudno's North Shore stared in amazement at the objects slowly and carefully picking their way through a small flotilla of pleasure boats out in the centre of the bay. The crowds had been growing for almost two hours ever since a shapeless amorphous mass had first rounded Little Orme's Head. As it turned beam on the silhouette of one of HM's armed trawlers, BCK912, presented itself to the telescopes of the curious. But what had it got in tow? Suddenly it dawned on them — it was an airship! But one such as few had not seen before, and they had seen all sorts over the last thirty months as naval blimps hugged the Denbighshire coastline returning to station. To be more precise it was the undignified hulk of SSZ.35, a non-rigid airship of the Sea Scout 'Zero' type, based at Llangefni Naval Airship Patrol Station, RNAS, Anglesey. The boat shaped aluminium car occasionally crested a wave on the otherwise calm sea and a triangular 'forced landing' pennant dangled forlornly from its halliard forward. Above it, still slowly deflating, the 143 ft. envelope was assuming a weird and wonderful shape, threatening almost to jack-knife as, with pusher engine stopped, its air ballonets lost pressure.

It seemed as if the trawler might be heading for the lifeboat slipway to shed its tow, but this was lined with queues of people eagerly seeking to board the numerous launches moored to it at the water's edge. A close-up inspection of a crippled airship would add a new dimension to a pleasure cruise around Llandudno Bay. But the trawler hove to off the end of the pier where a group of soldiers were waiting. Disciplined hands took over the trailing rope and to the shouted commands of F/Lt. T. B. Williams (airship captain) SSZ.35 was slowly walked ashore and secured on a small grassed area adjacent to the

SSZ35 moored at Llandudno after engine failure. [ARO]

Hydropathic Hotel, there to await the arrival of the repair/salvage team from Llangefni, which, judging from the swelling sound of hooters, was not far away.

The present emergency had started thirty-two miles away off the Formby lightship which marked the seaward end of the Crosby Channel, the main access point to the port of Liverpool. Four Llangefni 'Zeros' were on patrol at the time with SSZ.35 more particularly detailed to cover 'the Liverpool Bay trade route', that is, escorting an inward bound convoy before returning to pick up a troopship sailing between Holyhead and Kingstown (Dún Laoghaire). The airship was working in close collaboration with surface vessels scouring Liverpool Bay for a reported enemy submarine when her Rolls Royce 75 h.p. Hawk engine packed up — just as a stream of oil and bubbles on the surface, moving steadily westwards, pin-pointed the exact position of the submarine! It was all the more frustrating for the airship's crew since the water cooled Hawk had been specially designed for the 'Zeros' — adapted aeroplane engines had consistently shown their unsuitability for slow flight — and, having overcome the usual teething problems, had gained a reputation amongst airship crews as 'the sweetest engine ever run — it only stops when switched off or out of petrol', but obviously not on this occasion! It was all the more worrying since SSZ.35 was a fairly new ship, trialled on 23 March 1918 and taken on strength the next day at Llangefni. It had made its first patrol on 8 April and engine-wise had only 153 hours flying time 'on the clock'.

The police had great difficulty in keeping back the curious crowds. Flight Engineer and Observer had perforce to stay with the car, attempting to sort out the engine — a faulty water pump was immediately suspect — while a frustrated captain was only slightly mollified by being dined out — after a bath and borrowing a tie — as guest of the The Hydro's' manageress. The repair squad, under the Station Engineer Officer, duly arrived complete with hydrogen bottles. An area was roped off and the unheeding and uncomprehending crowd warned of the fire hazard from hydrogen and petrol present in such large quantities. As a precaution the ship was further secured by lines attached to great iron spikes driven into the turf and a long linen hose run across to the air-scoop from a portable compressor to 'pressure up' the ballonets and help trim the vessel. Problems of filling the high fuel tanks slung beneath the envelope were solved with the co-operation of the local fire brigade who loaned their turn-table ladder. At 20.30 hrs. SSZ.35, now conforming more to the popular conception of a non-rigid airship, was able to limp back to station — only to experience further engine failure the following evening, just ten minutes into its test flight!

This was not an isolated incident along the North Wales coast. Early 1915 had seen the compulsory requisitioning of farmland in Anglesey, the establishment of a Royal Naval Air Station (RNAS) at Llangefni, and the intrusion of a largely English military presence into a virtually monoglot Welsh-speaking area. The arrival of submarine chasing blimps or non-rigid airships and their occasional enforced landings amidst unsuspecting rural communities called upon to give succour to these beasts of the sky — or even help walk them home! — was the region's introduction to war in the air. Two years later, as operational and training aerodromes were opened at Hooton Park, Shotwick (Sealand), Tern Hill and Shawbury and aeroplanes also started to fall from the sky, their induction would be complete.

Llangefni RNAS was commissioned on 26 September 1915, part of a chain of 'war stations' around the coast of Britain established to combat the increasing menace of enemy submarines. In Llangefni's case it was, along with Pembroke (commissioned January 1916) in the south and Luce Bay, Wigtown (opened 15 July 1915), in the north, to provide air cover over the Irish Sea with its 'trade routes' into Liverpool and the Holyhead mail boat and troopship links to Ireland.

To this end the 'low and slow' and hovering characteristics of the

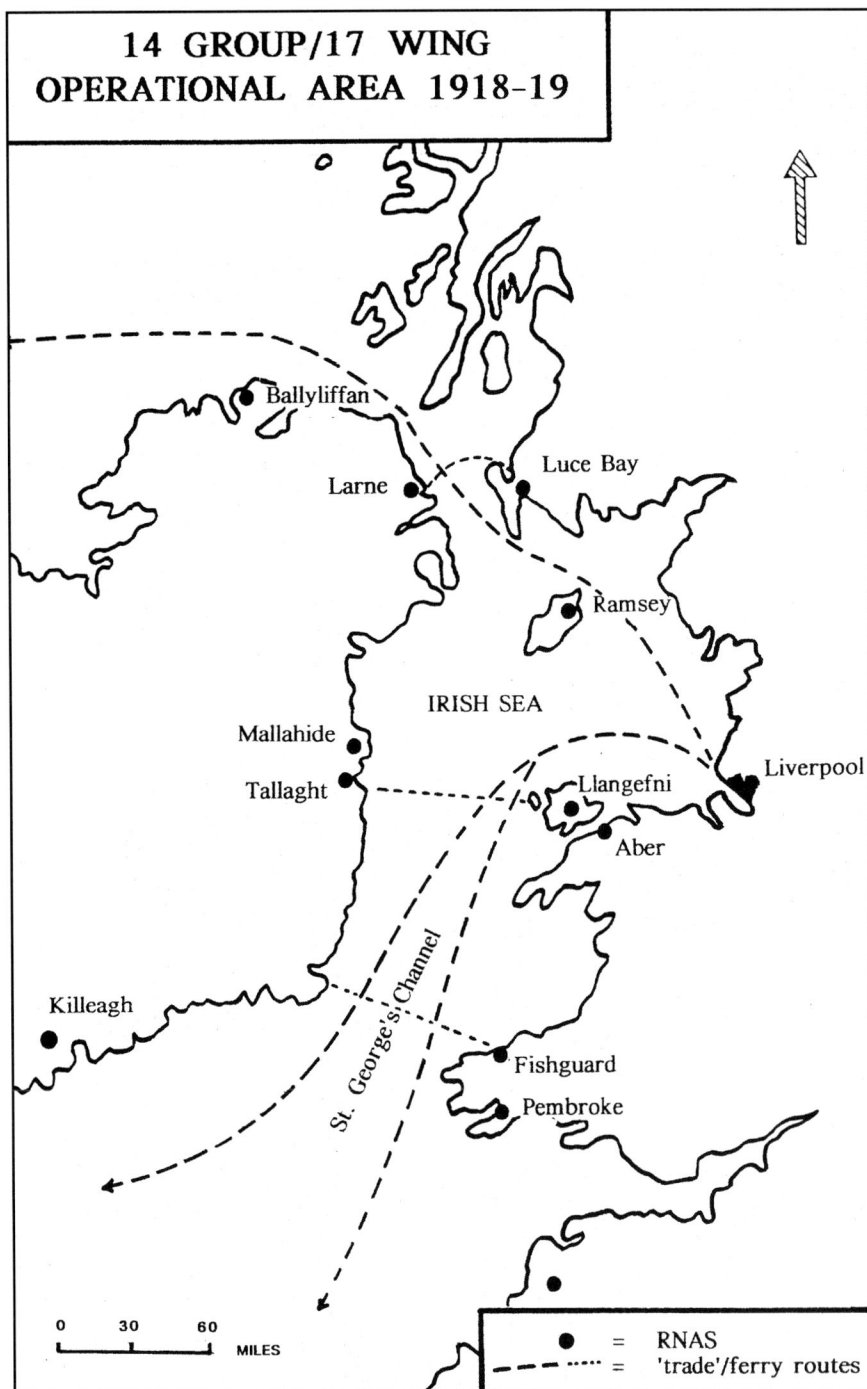

14 GROUP/17 WING OPERATIONAL AREA 1918–19

IRISH SEA

Ballyliffan
Larne
Luce Bay
Ramsey
Mallahide
Tallaght
Llangefni
Liverpool
Aber
Killeagh
Fishguard
Pembroke

St. George's Channel

0   30   60   MILES

● = RNAS
— — — ···· = 'trade'/ferry routes

would be forced to remain submerged, thus restricting its speed, adversely affecting attack strategy, and hopefully allowing merchant ships to outpace the would-be attacker. Should the track of a U-boat be found, or if one was foolish enough to surface or attack while an airship was in the area, a radio call to the nearest destroyer or armed surface vessel brought immediate assistance and retribution.

In its early days NAPS Llangefni was also referred to as NAPS Bodffordd, Gwalchmai or Heneglwys after nearby hamlets, all equally unpronounceable to the average Englishman, hence the compromise name, NAPS Anglesey! The station was charged with the policing of a huge sea area Bardsey–Dublin–Isle of Man-Barrow–Point of Ayr (Flintshire). It was situated on the north side of the A5, three miles west of Llangefni, and extended overall to some 260 acres, bounded on the north and east by the roads to Heneglwys. Dislocation of civilians was minimal. Indeed small holdings within the notional perimeter such as Tros-y-rhos ('across the moor'), the latter almost on top of the Silicol gas plant and hydrogen storage tanks, were left intact and the fields around the boundary grazed and mown although not ploughed. After stand-down in January 1919 the station went onto a 'care and maintenance' basis for two years before the technical/accommodation site was sold to Anglesey County Council for a hutted hospital and a County Surveyor's depot. In 1942 the site was re-requisitioned by the Air Ministry and roads diverted for the building of RAF Mona (initially RAF Heneglwys, but still no easier to pronounce!), home to the Bothas and Ansons of 3 AGS. The aerodrome is still used as a RLG by units operating out of RAF Valley.

Staffed by some 130–150 officers and men and thirty civilians, NAPS Llangefni was initially commanded by Major G. H. Scott (September 1915-October 1916>). He was followed by S/Cdr. A. Corbett Wilson (<May 1917>), S/Cdr. Brotherton (<March 1918>) and Major (later Air Marshal) Thomas Elmhirst (<November 1918-January 1919). Scott went on to greater things, eventually becoming Deputy-Director of Airship Development. In April 1918 he was captain of the wartime rigid experimental/training airship HMA No.25r at Cranwell, while in 1919 he commanded the R. 34 on its Baltic 'cruise' (June) and its double trans-Atlantic crossing (July). On 5 April 1921 he was commanding R. 36 (civil reg. G-FFAF) on an eventful PR flight in which everything went wrong! Tragically he would lose his life in the R101 disaster. A regular Naval officer who had seen action at the Dardanelles and Dogger Bank in HMS *Indomitable*, Major Elmhirst had transferred to the RNAS in 1915, cutting his teeth as captain of the SS.17 out of Luce Bay in 1915, captain of a 'Coastal' class type (C.19) in 1916, and of HMA No. 5, one of the huge Vickers-built 'Parsevals' at Howden, in 1917. His stint as CO Anglesey earned him the AFC. He was Scott's navigating officer on the R. 36 in 1921 and in 1925 passed through RAF Staff College. He commanded 15 (Bomber) Squadron at Abingdon 1935–7. During the Battle of Britain, Air Cdre. Elmhirst was Deputy Director of Intelligence at the Air Ministry, going on to make his mark as AOC Egypt in 1941 and 2i/c Desert Air Force during the Alamein campaigns. He returned to NW Europe in 1944 as 2i/c British Air

non-rigid airship proved ideal as they scoured the sea for periscope wake, underwater shadow, or floating mine. They easily outstripped contemporary aircraft in terms of range, endurance and carrying capacity. However, from March 1918 onwards, as U-boat activity intensified, especially in coastal waters, the pressure on airships would be eased somewhat by the introduction of special coastal patrol or 'scarecrow' flights using surplus Airco D.H.6 trainer aeroplanes for the purpose.

The destruction of submarines from the air was a matter of great skill and practice. By contrast frightening them off from potential targets was a relatively simple matter, achievable with easily trained volunteer crews, leavened with a lump of Western Front veterans desperate to get back into operational flying. The principle behind 'scarecrow' patrols — both blimps and aeroplane — was that a submarine would spot the airship long before the blimp could sight the U-boat. The submarine

Forces, Normandy-Germany, and was the first C-in-C Indian Air Force 1947-50, when he retired. But from his *Recollections* it is clear that he looked back upon his airship days with nostalgia and a particular fondness.

~~The technical and accommodation~~

14 GROUP/17 WING OPERATIONAL AREA 1918-19

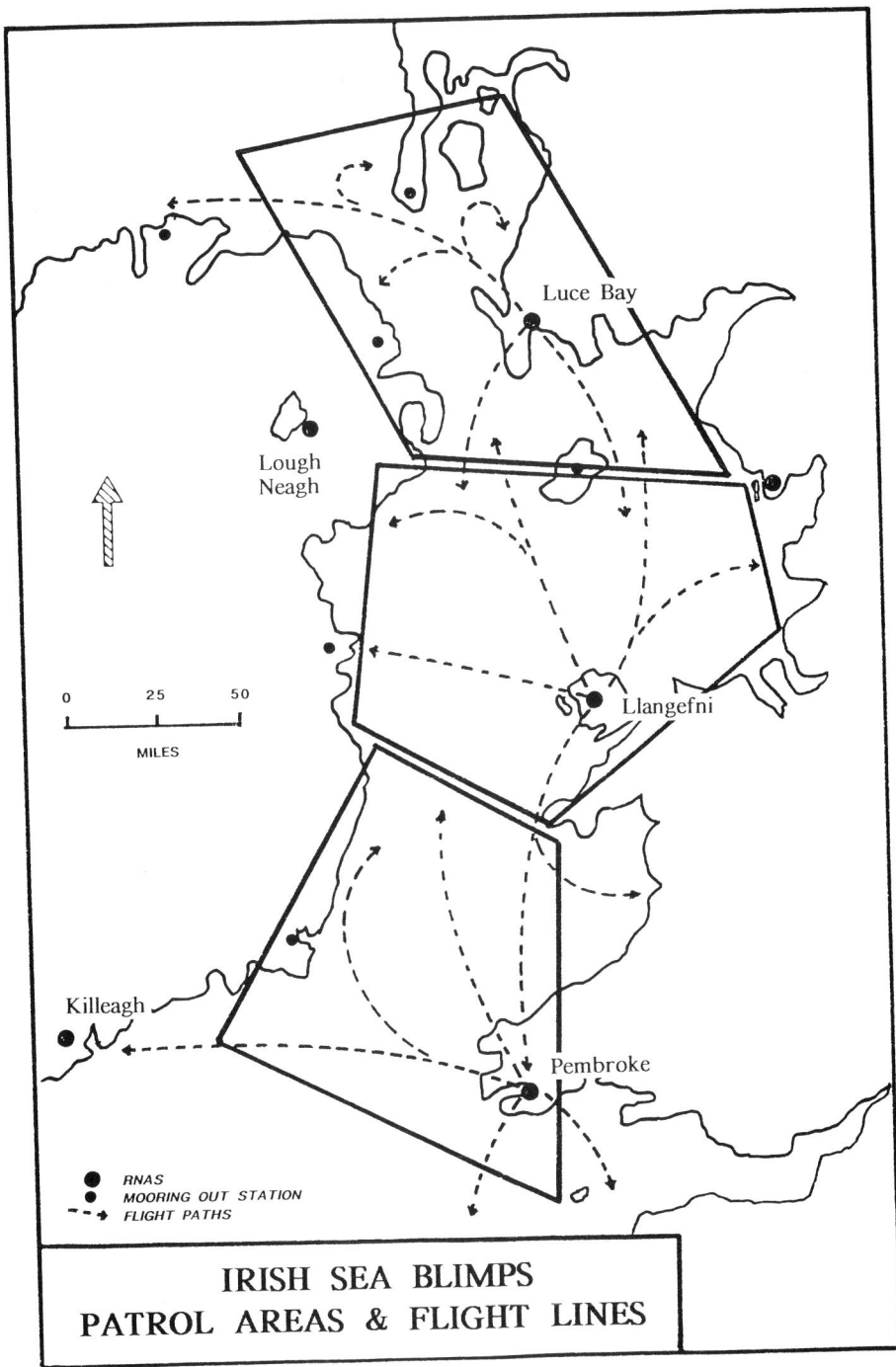

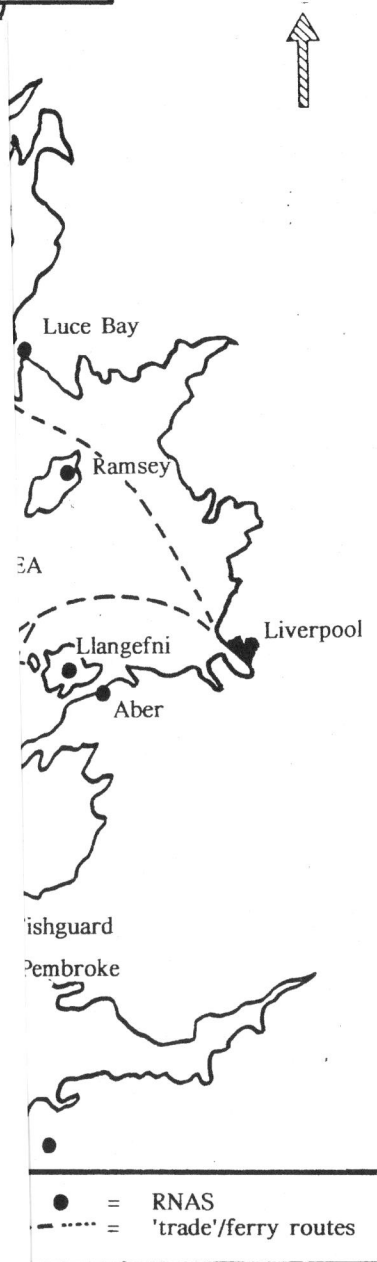

IRISH SEA BLIMPS PATROL AREAS & FLIGHT LINES

Legend:
● = RNAS
● = MOORING OUT STATION
- - → = FLIGHT PATHS

● = RNAS
- - - - - = 'trade'/ferry routes

0   25   50
MILES

CORRECTION: Due to a printing error, the incorrect map was placed on page 39. The problem has been resolved by the insertion of this page, showing the correct map: 'Irish Sea Blimps Patrol Areas & Flight Lines'

...ngefni's 'mooring out' station in the ...Dublin, principal seat of Lord Talbot. ...ment of soldiers that, after training, ...king out airships! It had long been ...Ireland's east coast would ease the ...s and reduce the unreasonably long ...me Pembroke RNAS was provided ...h (Co. Cork). To the north Luce Bay ...ing-out facilities at Ramsey (Isle of ...an (Donegal). Ultimately Malahide ...ex-Army portable airship sheds developed in anticipation of mobile warfare, consisting of a steel framework covered by camouflaged canvas. The wedge-shaped doors just enabled two 'SSZ' ships to fit in nicely. But until this was ready airships were moored in a clearing amongst the trees. A pit was dug to

...Duties of 285 Squadron, to upgrade the surface for regular use by the Special Duties Flights would have been prohibitive, hence their early transfer to an aerodrome on the Welsh mainland between Penrhyn Castle and Abergwyngregyn, 'Aber' for short.

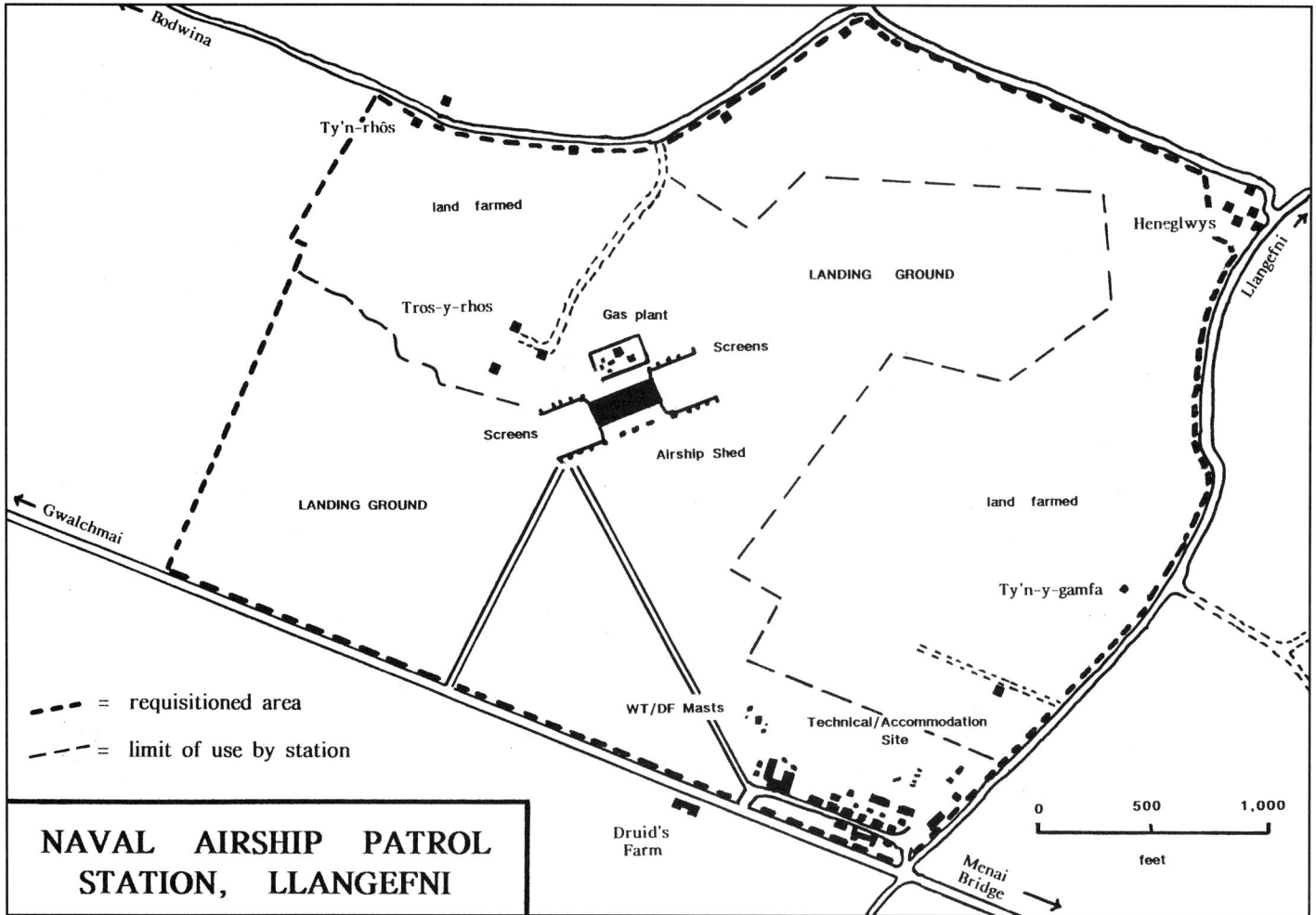

NAVAL AIRSHIP PATROL STATION, LLANGEFNI

Bodwina

Ty'n-rhôs

land farmed

Tros-y-rhos

Gas plant

Screens

Screens

Airship Shed

LANDING GROUND

Heneglwys

Llangefni

LANDING GROUND

land farmed

Ty'n-y-gamfa

Gwalchmai

WT/DF Masts

Technical/Accommodation Site

= requisitioned area

= limit of use by station

Druid's Farm

Menai Bridge

0   500   1,000

feet

The Naval Airship Patrol Station at Llangefni from the south-east. [ARO]

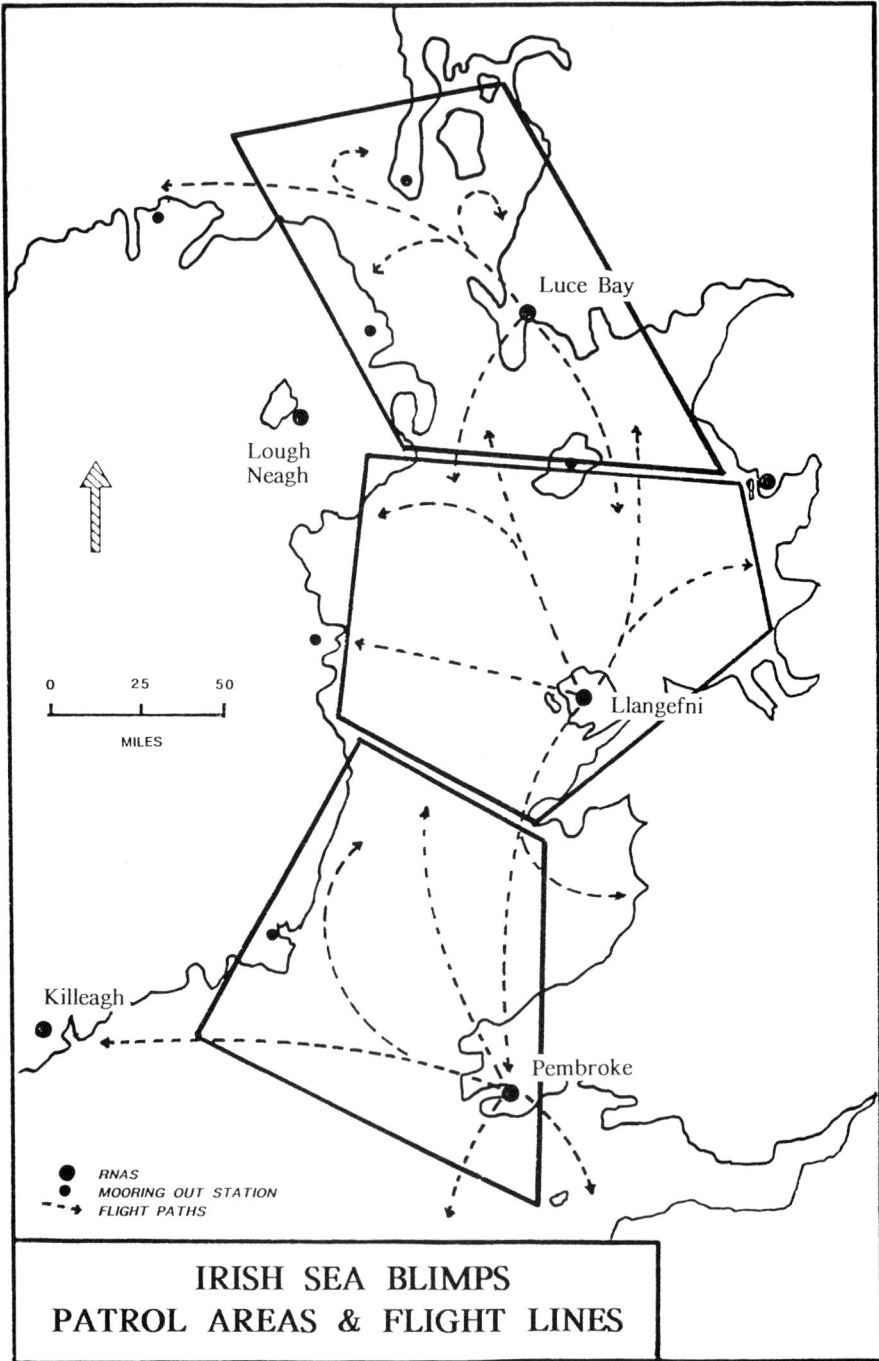

Luce Bay

Lough
Neagh

0   25   50
MILES

Llangefni

Killeagh

Pembroke

● RNAS
● MOORING OUT STATION
- - → FLIGHT PATHS

IRISH SEA BLIMPS
PATROL AREAS & FLIGHT LINES

CORRECTION: *Due to a printing error, the incorrect map was placed on page 39. The problem has been resolved by the insertion of this page, showing the correct map:*
'Irish Sea Blimps Patrol Areas & Flight Lines'

Forces, Normandy-Germany, and was the first C-in-C Indian Air Force 1947-50, when he retired. But from his *Recollections* it is clear that he looked back upon his airship days with nostalgia and a particular fondness.

The technical and accommodation site was tucked away in the south-east corner of the station, running parallel to the A5. The main gate was just past Druid's Farm and gave access to picket post, guard room, station HQ and meteorological hut. WT hut and masts were set back on their own. One tends to forget that although in its infancy, of short range and often unreliable, radio was vital to the airship service and by March 1916 a Marconi relay transmitting and receiving station had been built at Amlwch and a D/F station on Colwyn Bay Golf Club. Past the flag-staff in its little grassed plot defined by flower beds and the ubiquitous white-painted stones, stood the water tower and well, the station chapel, NCOs' huts and barrack blocks behind. Further back there were two huts for officers. At the cross-roads stood the Regimental Institute building, and along the Heneglwys road MT shed, stores, armoury, power house and, well out in the field, the bomb dump.

In February 1915 the Admiralty produced its specification for a new emergency class of airship — small, 50 m.p.h. maximum speed, minimum 8 hours endurance, carrying a crew of two, 160lbs of bombs and a W/T set with a radius of 30 miles so that surface ships could be called up to deal with submarines as sighted. The result was the 'SS Class' ('Sea Scout' or 'Submarine Scout'). As produced the airship was 143ft long, 27ft in beam and 60,000 cu. ft. (gas capacity), with, as an expediency, an existing aeroplane fuselage minus wings and tail unit, adapted as the car — a Bristol BE2c fighter (with tractor engine) or a Maurice Farman (with pusher engine) according to which company built the airship, RNAS Kingsnorth or Airco. Pusher engines were most favoured by airship crews largely because they did not have to sit in the slipstream!

To house the Sea Scouts at Llangefni one large airship shed, 323ft x 120ft 5ins x 80ft with huge sliding doors at either end, was erected in the centre of the station aligned NE-SW into the prevailing wind. It was built by specialists from the Air Construction Corps and could comfortably house four inflated 'SS' class blimps. The huge shed was essential. The extreme buoyancy of airships meant that they needed careful handling when on the ground, especially in bad weather. Even Llangefni's small blimps needed 30-40 men to handle them in a strong breeze. Only when inside a shed was an airship safe. Two pairs of screens as long as the shed extended out at each end into the landing areas. These protected the blimps from sudden gusts of wind as they were 'walked out' by their ground crews. The landing areas were the only part of the station to have their hedges removed and ditches filled. It was rough 'moorish' ground so that although the station initially maintained a Bristol Scout D as a communications aircraft and for a short while was home to seven Airco DH6s of 255 Squadron, to upgrade the surface for regular use by the Special Duties Flights would have been prohibitive, hence their early transfer to an aerodrome on the Welsh mainland between Penrhyn Castle and Abergwyngregyn, 'Aber' for short.

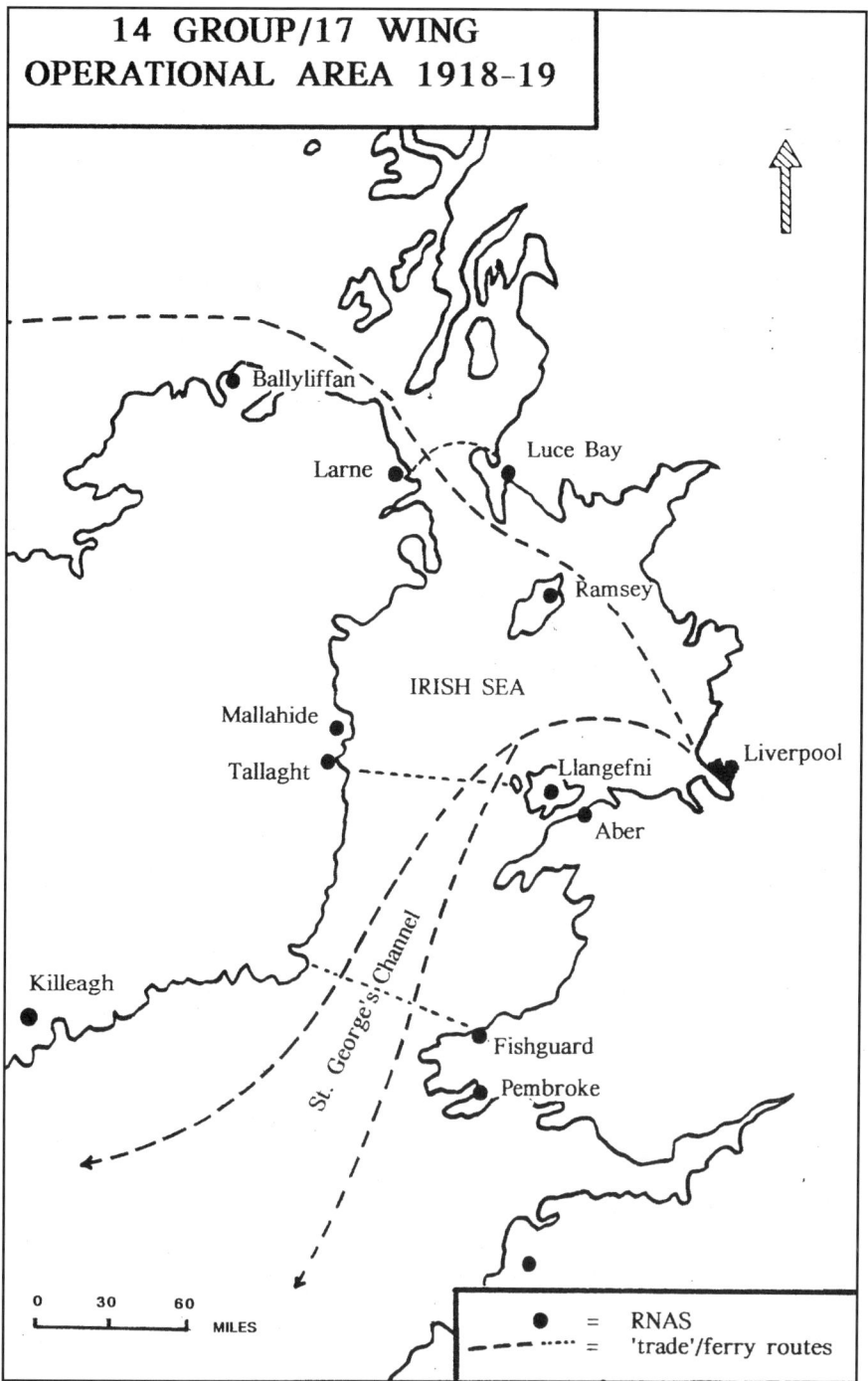

14 GROUP/17 WING OPERATIONAL AREA 1918-19

RNAS = ●
'trade'/ferry routes = – – – ·····

In June 1917 work started on Llangefni's 'mooring out' station in the grounds of Malahide Castle, near Dublin, principal seat of Lord Talbot. This had the advantage of a detachment of soldiers that, after training, was roped in for landing and walking out airships! It had long been apparent that an airship station on Ireland's east coast would ease the pressure on Llangefni-based blimps and reduce the unreasonably long hours spent at sea. At the same time Pembroke RNAS was provided with a subsidiary station at Killeagh (Co. Cork). To the north Luce Bay would also be furnished with mooring-out facilities at Ramsey (Isle of Man), Larne (Antrim) and Ballyliffan (Donegal). Ultimately Malahide was provided with one of the ex-Army portable airship sheds developed in anticipation of mobile warfare, consisting of a steel framework covered by camouflaged canvas. The wedge-shaped doors just enabled two 'SSZ' ships to fit in nicely. But until this was ready airships were moored in a clearing amongst the trees. A pit was dug to

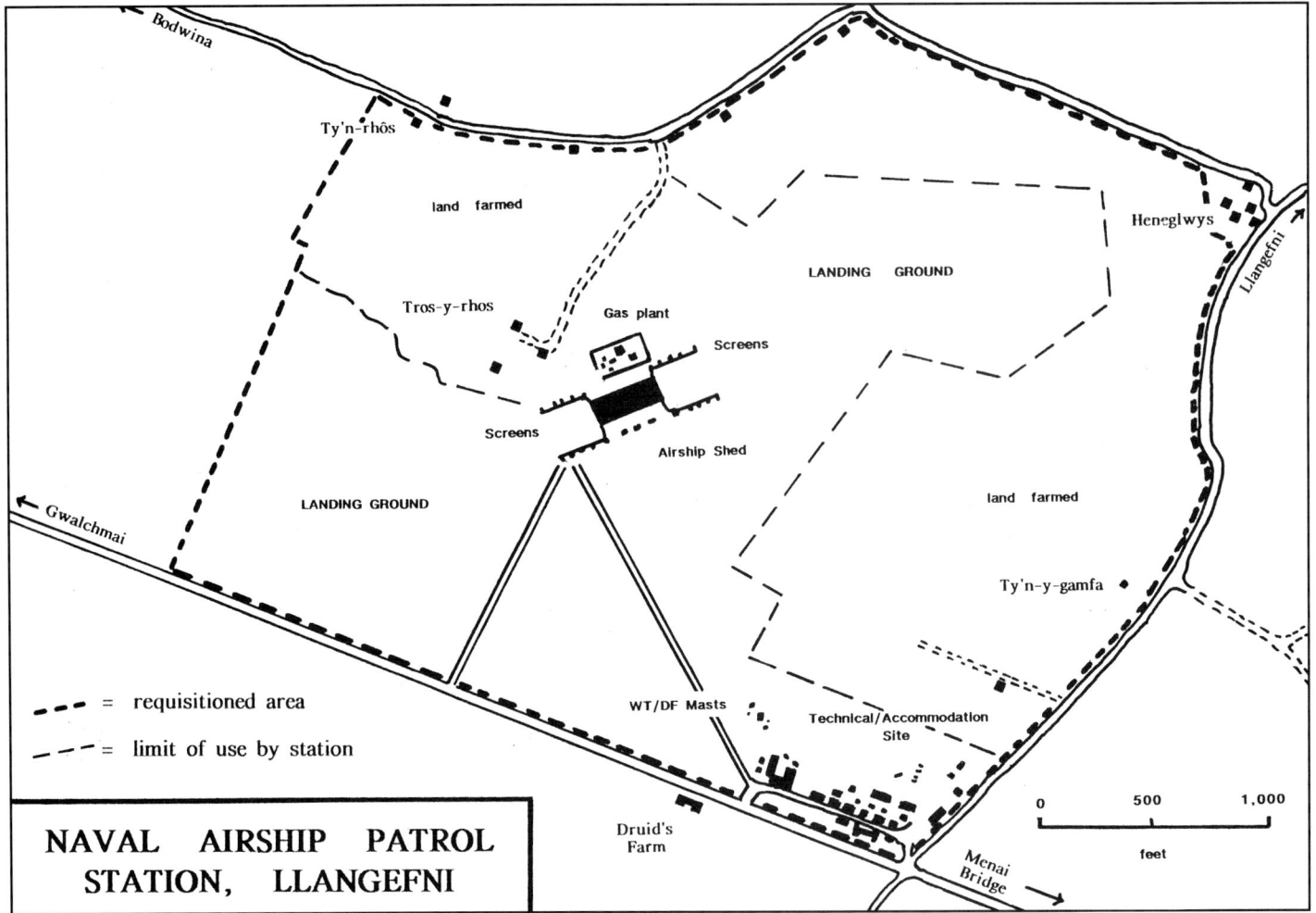

NAVAL AIRSHIP PATROL
STATION, LLANGEFNI

- - - = requisitioned area

- - = limit of use by station

*The Naval Airship Patrol Station at
Llangefni from the south-east.*
[ARO]

accommodate the car, bringing the envelope closer to the ground for easier maintenance and greater protection from the elements. Accommodation was tented and ground affairs were managed by a Chief Petty Officer from Anglesey. Malahide became fully operational early in March 1918 but it was not a pleasant detachment for Llangefni men — the memories of the Easter Uprising of 1916 and its savage repression were still fresh in the minds of Irish folk!

The first blimp to arrive at Llangefni was SS.18, crewed by FSLs Urquart and Kilburn. It flew in to Anglesey from Kingsnorth on 26 September 1915, the station's commissioning date. It had a BE2c car with additional long-range fuel tanks fitted beneath the fuselage, bomb racks aft of the undercarriage, grappling iron and trail ropes, and 300lbs (30 gallons) of water ballast. The W/T operator/observer occupied the front cockpit and pilot the rear (order reversed in the Maurice Farman cars). The two ballonets of 6,500 cu. ft., one aft., one forward, and essential for maintaining the vessel's shape and trim, were inflated from a metal scoop mounted in the propeller's slip stream. By 5 November 1915 SS.22 (FSL F.E.Turner), SS.24 (FSL Scroggs) and SS.25 (FSL T. B. Williams) had arrived to complete the original establishment.

The young Flight Sub-Lieutenants had just passed out after 'Balloon Training' at Wormwood Scrubs. Their 'Sea Scout' mounts were to give them first class experience and prepare them for command of the larger, improved airships that were to appear in the ensuing years. Kilburn would be seen in 1917 as a Flight Lieutenant, captain of 'Coastal' type C.26 based at Pulham, Norfolk, thrashing his ship to the limit of its endurance, running out of fuel and drifting over Holland where he and his crew were interned! FSL T. B. Williams would spend much of his wartime career at Llangefni. After delivering SS.25 his first command was SS.31, the experimental 'Flying Bedstead', at Kingsnorth, September–December 1916. He would trial SSP.6 on 8 June 1917 and fly it to Anglesey eight days later. In March October 1918 Williams would command SSZ.35 at Llangefni, quite an eventful seven months! After losing SSZ.35 at sea on 17 October 1918, Williams acted as chief officer on the disaster prone 3¹/2-day flight (28-31 October 1918) from Rome to Kingsnorth of the Italian 'M' class airship, SR-1. Between February–October 1920 he had command of SSE.3 ('Experimental') at Howden, training rigid airship crews. This was the last Naval non-rigid to fly.

Weather and serviceability permitting, the 'Sea Scouts' were seldom off patrol. Dispatched in relays from first light they patrolled from eight to ten hours at a time, hard going, especially for the pilots, watching course, pressure and scouring the surface of the sea for traces of the invisible enemy. Elevator wheel controls raised blisters on soft hands, especially on a 'bumpy day'. Cramped backsides, frozen feet, bursting bladders, gnawing pangs of hunger — one could never eat or drink properly in the slip stream of an 75 h.p. open exhaust Renault engine — few scarecrows in a farmer's field had to put up with the conditions under which the 'animated scarecrows' of the sky fought their own little bit of the war. If at the end of a patrol pilot and observer had to be lifted bodily out of their seats before circulation could be restored, the winter of 1915–16 also had its positive side. Submarine activity was minimal so that airship crews could settle in and perfect their handling skills, especially in bombing and machine-gunning, using for the purpose — stray sheep permitting — the full-size dummy U-boat target marked out in one corner of the Llangefni landing ground.

The Llangefni 'Sea Scouts' had a fairly good safety record. But accidents were bound to happen, particularly in forced landings as the Renault engines overheated and failed — they were designed to drive lighter craft at greater speeds through the sky. Thus, on 23 November 1917 SS.25, despite grapnels and trailing ropes being let down, landed heavily, crashing through some tree tops before wobbling to a halt, damaging car and elevator planes. In that it retained the same

Right: FSL T. B. Williams flies SSP.6, grasping elevator with right hand. Instrumentation was minimal, compass centre top, aneroid and height indicator below, stop watch to left. [ARO]

registration number the damage was not all that serious. Airship serial numbers stayed with the car, regardless of the number of envelopes used. If the car was replaced an 'A' serial number would be raised.

As far as scant surviving records show, there was only one 'fatal' accident. On Sunday, 22 October 1916, SS.18, whilst attempting to land, struck a cow fatally injuring the animal and carrying away the undercarriage and damaging the car. By this time its BE2c fuselage car had been enlarged in the station workshops to take an extra cockpit cut out in the rear fuselage. The observer jumped, or was thrown, out of his forward cockpit. The injured airship, lightened by some 150lbs, sprang upwards out of control. Inexplicably the envelope was not 'ripped' to release gas and lose height. As a 'free balloon' SS.18 was carried out to sea. When at last the pilot, FSL Arthur Donald Thompson, managed to valve some gas, the airship fell heavily into the water, rupturing the envelope and splitting the car in two, throwing out the flight engineer, who, weighed down by heavy flying gear, was unfortunately drowned. FSL Thompson managed to cling to the wreckage before being picked up by a passing ship. Obviously attributing his survival to Divine Providence, he later presented two candle-sticks to St. Cyngar's Church, Llangefni.

SS.18 was officially written off on 9 November 1916, but had been replaced three days earlier by SS.33. The latter carried a Maurice Farman fuselage type car with a Rolls Royce 'Hawk' pusher engine. In April 1917 she in turn suffered engine failure returning from patrol and made a forced landing in Cemlyn Bay, but with only slight damage. On 9 May, she was back in harness, working along with M.L. 221, an armed launch out of Holyhead, running the first of a series of successful air-to water trials on the new-fangled ASDIC (Anti-Submarine Detection Investigation Committee), an early form of sonar, which, along with the depth charge, would mean a quick end for any detected U-boat. Airships hovering or moving 'dead slow' to reduce microphone 'noise' were ideal for the purpose, drawing a 'hydrophone' on the end of an 180ft cable through thirty foot of water. Experiments would continue into 1918, but the war had ended before the use of the device was sanctioned operationally. The problem with using 'SS' type blimps for the purpose was purely a logistical one — the storage and winding of some 65 fathoms of high tensile steel cable in a BE2c car!

The 'Sea Scout' class of airship, some forty-nine in all, had been hastily produced to meet a particular deficiency at a particular time; they were little more than stop-gap craft., with glaring limitations performance-wise. In the Admiralty's scheme of things they would be

Llangefni's D/F station on the Golf Course at Bryn-y-maen, above Old Colwyn. [DRO]

Left: An unhappier moment. SS.25 towed into Holyhead by the armed yacht Amethyst *after engine failure. [ARO]*

Ground crew take over SSZ.35 as she comes in to land at Llangefni. [ARO]

Left: FSL T. B. Williams' first command, SS.31, the 'flying bedstead', with modified Farman 'pusher' car.

replaced by a larger airship with greater range and a reliability enhanced by twin engines. Like the 'SS' class, the new 'Coastal' types were also designed around the fuselage of obsolescent aircraft, if for no other reason than there was a plentiful supply of the things and they could be mass-produced. The 'Coastal' car utilised two Avro 510 seaplanes, tails shorn off aft of the cockpit and stuck back to back as it were, with the 150 h.p. Sunbeam 'Crusader' engines forward and aft, the latter obviously with a 'pusher' screw. The 'Coastals', 195ft long, of 170,000 cu. ft. capacity in a distinctive tri-lobe envelope, carried a crew of four, coxwain, pilot, W/T operator and engineer, although at a push an extra observer/passenger could be fitted in amidships in place of the water ballast bag! Endurance was 20hrs, maximum speed 45 m.p.h.. They were armed with two Lewis guns and in practice could carry three or four times their specified four 100lb bombs. Although forty-two 'Coastals' of various types were built, none were allocated to Llangefni. Three were allocated to RNAS Pembroke — C.13, C.5A and C.6. In that Pembroke patrolled the southern part of the Irish Sea in a quadrant St. David's Head–Bardsey–Black Rock (Co. Wicklow)–Tuscan Rock (Co. Wexford) Lundy, its 'Coastals' regularly dropped in at Llangefni for emergency repairs, refuelling or if caught out by bad weather. More frequently it was after handing over a convoy they had picked up at Wexford to Llangefni's blimps off the Skerries and then literally having time on their hands before escorting another convoy ex-Liverpool southwards. Conversely Llangefni's blimps would regularly make longer trips south, zig-zagging over the Cardigan Bay coastal shipping routes and landing at Pembroke to refuel, the round trip taking some fourteen hours.

Instead of 'Coastals' Llangefni received in 1917 the 'SSP' class type ('Sea Scout Pusher' or Submarine Scout Patrol'), 143ft long, 30ft in diameter, capacity 70,000 cu. ft. Only six were built and three of those operated out of Anglesey. The Llangefni ships, SSP.1, SSP.5 and SSP.6 would serve until after the Armistice. Of the other three, two were lost at sea before the year was out and one was wrecked on its delivery flight! Since the most popular 'SS' airships were those with the Maurice Farman car and pusher engine, the new type also had a specially designed longer car with rear pusher engine. It carried a crew of three — pilot, W/T operator and engineer.

Delivery of the 'SSPs' to Llangefni was staggered over six months, beginning with SSP.5 on 18 January 1917. SSP.6 arrived on 16 June 1917, followed by SSP.1 on 5 July. This enabled crews to familiarise themselves with the new airships, new crews to be trained where necessary and at the same time the increasing demands for patrol and escort blimps to be met. There was an immediate problem with airship accommodation as the 'SS' ships were not dispensed with immediately. One can only assume that some were farmed out to Malahide even though that mooring-out station was no yet fully completed.

The first 'Sea Scout' to go was SS.22 which went to Wormwood Scrubs in April 1917, whence she embarked upon an even more eventful career under somewhat erratic Italian management. SS.24 was transferred to Luce Bay in the July and would have another year's hard wear and tear before being written off as unserviceable. SS.25 and SS.33 would soldier on at Llangefni until March 1918. When finally replaced by the new 'Zero' type SSZ.34, SS.25 had logged over 843 flying hours and had remained inflated even longer. SS.33 was replaced by 'Zero' SSZ.35.

Had plans matured SSP.5 would never have reached Anglesey. Originally built at Wormwood Scrubs, it had been flown to Kingsnorth and fitted with Remy ignition gear, a silent engine and a black envelope (85,000 cu. ft) for secret night flying with the BEF in France. The project was cancelled and SSP.25 returned to Wormwood Scrubs for 'normalisation' and thence to Anglesey. Whilst at Llangefni (18 January 1917–26 march 1918) it notched up 370 flying hours.

Over its thirteen months on the island SSP.6 logged 320 flying hours.

A Pembroke-based 'Coastal' type (C3) calls in at Llangefni in 1917 at the end of a 'long leg' over Cardigan Bay. [ARO]

To it, on Friday, 16 March 1918, is attributed one of the most bizarre incidents that ever befell an Anglesey blimp and its crew. It had come down in the Irish Sea with engine trouble — the 100 h.p. water-cooled Greens were just as unreliable as the early Hawks. As the crew were taken off by a passing ship and preparations made to try and take the blimp in tow, SSP.6 rose, took on the aspects of a free balloon and floated away south-eastwards. Its erratic course was plotted across nine counties before it eventually landed in a wood at Chichester (Sussex) where it was taken in hand by a puzzled garrison until relieved by an equally perplexed salvage crew from the RNAS sub-station at Slindon, some six miles away. Her temperamental displays were not yet over. On 12 August 1918 SSP.6 left Llangefni for her training role at HMS *Daedalus*, otherwise the Airship Training Wing at RNAS Cranwell, Lincs. Piloted by FSL G. E. Bungey, she had to make a forced landing near Blackburn due to engine failure (still that 100 h.p. Green!) and was badly damaged.

The proud boast of the RNAS was that no ship was ever sunk if it was being escorted by an airship. It must have been frustrating for the Llangefni crews to learn of the many sinkings off Anglesey and in Liverpool Bay in the latter half of 1917. In these coastal waters ships often elected to sail alone. As the 'convoy system' drastically reduced shipping losses, these 'loners' provided frustrated U-boat commanders with attractive alternative targets. In October 1917 Llangefni had SS.25, SS.33 and SSPs. 1, 5 and 6 operational, working flat out, mainly on deep water patrols and escorts. Despite their comparatively limited endurance, aircraft were desperately needed to patrol within a few miles of the coastline where the majority of opportunists sinkings had occurred. The Admiralty responded positively, but it was a piecemeal reaction to requests from individual war stations. As yet there was no co-ordinated anti-submarine strategy.

On Wednesday, 7 November 1917, six Airco DH4s left Hendon (later No. 2) Air Acceptance Park for Llangefni. These were two-seater day bombers, powered by a 375 h.p. Rolls Royce 'Eagle' engine giving a maximum speed of 136 m.p.h. and an endurance of some three hours. It could carry a 460lb bomb load on external racks and was armed with two Lewis machine guns, one synchro forward firing and one in the observer's cockpit. One might think ideal aircraft for the task ahead. But the outcome was shambolic. Over the Dee estuary the aircraft ran into bad weather strong winds, driving rain, low cloud and mist, which made flying hazardous. Four aircraft turned back and sought refuge at the newly opened Shotwick aerodrome (otherwise Sealand). Two pressed on, one making a discretionary landing on the exposed Lavan Sands (Traeth Lafan) off Abergwyngregyn. Despite efforts to tow it to

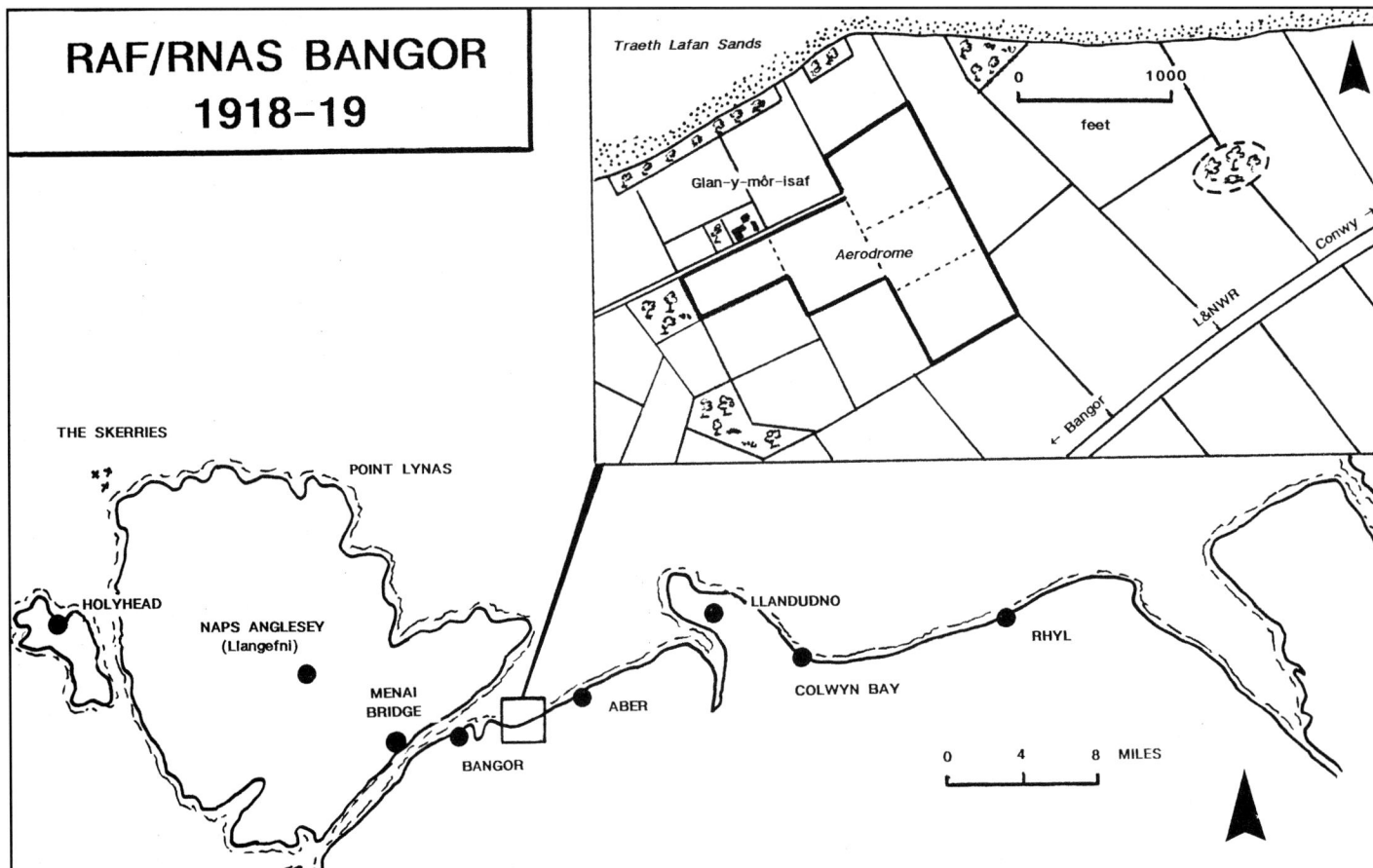

RAF/RNAS BANGOR 1918-19

safety with farm horses it was lost to the incoming tide. The second, A7654, actually made it to Llangefni, but crashed during the landing attempt. It was caught by a strong gust of wind as banking and side slipped in, striking a stone wall and killing the pilot 2/Lt. Carter. The observer, Cpl. H. Smith, was seriously injured.

Although over the next four months some twenty-nine more ships would be sunk in Llangefni's patrol area with the loss of over fifty lives, the DH4s did not return. Their replacement in the Summer of 1918 comprised obsolete and surplus Airco DH6 two-seater trainers adapted as light bombers for an anti-submarine role. With their 90 h.p. RAF 1a, 80 h.p. Renault or 90 h.p. Curtiss OX-5 engines they were capable of 75 m.p.h. maximum and could carry only half the bomb load of a DH4. But they were supplied not at the request of individual stations but as part of a wider package as a new anti-submarine strategy was adopted, crucial to which was the philosophy that, since the likelihood of the destruction of an enemy submarine by an aeroplane was remote, 'scarecrow' patrols did not need front-line aircraft or the most experienced crews. Obsolete aircraft., readily available in vast quantities, manned by second-rate crews, could supply the necessary deterrent factor equally effectively and at half the cost!

In this context one should note two events of import that occurred early in 1918. In their struggle against the U-boat the Admiralty began to implement a policy of 'combined operations'. Llangefni's airships had long co-operated with Royal Navy surface vessels but now the hunt was to be controlled and co-ordinated on an area basis. To this end Captain Gordon Campbell, V.C. (of Q-ship fame) was appointed supremo of anti-submarine forces, both sea and air, at Holyhead. Initially he had under his command a flotilla of superannuated destroyers, an armed steam yacht, an armed drifter which doubled as a harbour defence vessel, some twenty-two lightly armed motor launches, four airships (rising to six by the end of March) and (on paper at least) a squadron of aeroplanes! In May Campbell was able to

move his flag into the light cruiser HMS *Patrol*. Moves towards this state of affairs had been made in November 1916 with the creation of an Anti-Submarine Division with the Naval Staff. In April 1917 a new South-western Group took over responsibility for anti U-boat operations in the Western Approaches. As enemy submarine activity increased this was further sub-divided in November 1917 into a Milford Haven Group with responsibility for the Irish Sea, and in March 1918, as already noted, into a Holyhead Group.

Eventually the air element in the new strategy would be underwritten to the tune of 1,060 DH6s distributed between some thirty-four Special Duties Flights or Coastal Patrol Flights deployed around the more vulnerable parts of Britain's coastline. It was several weeks before any aircraft materialised in North Wales. On 14 March 1918 S/Cdr. J. W. K. Allsop, in company with Llangefni's CO, S/Cdr. Brotherton, toured Anglesey and adjacent parts of Caernarfonshire looking for a suitable aerodrome site. Despite the fact that four months earlier Llangefni was considered suitable to house DH4s, it was a different matter for the lighter DH6s. The airship landing grounds were considered too rough without a great deal of preparation at too great an expense. S/Cdr. Allsop had commanded 4 Squadron at Bray Dunes (Belgium) in February-March 1917 before taking over 7 Squadron at Coudekerque (April-June) and was therefore something of an expert in the establishment of 'forward' airfields and converting flights into squadrons.

In May 1918, some 50 acres of Glan-y-môr-isaf farm, on the coastal flat a mile east of Penrhyn Castle, was selected as a 'Special Duties Station' and duly commissioned (?25) July 1918 by the simple expedient of uprooting two hedges, erecting tent lines and four Bessoneau hangars, digging bomb dump trenches and commandeering a farm outbuilding for the W/T station. The first aircraft arrived on 14/15 August. But until then patrol aircraft had to use Llangefni landing ground. The only problem was the inability for the DH6 to carry a reasonable pay-load. This meant that they often flew with an observer and no bombs, ideal for convoy escort duty, where feeding information to surface vessels was of the essence. When the flight moved to Glan-y-mor-isaf and the number of aircraft doubled, three out of four DH6s were flown as single seaters — the only way they could carry an adequate bomb load!

But before Captain Campbell saw any of his aeroplanes the Admiralty's new initiative was hi-jacked and modified by the merger of the RNAS and RFC on 1 April 1918 to form the RAF. The immediate impact was to impose upon a loose Naval organisation the RAF's hierarchal orthodoxy of Group, Wing, Squadron, Flight. Whether they liked it or not the RAF was committed to the Admiralty's 'scarecrow' expedient using DH6s.

No. 14 (Marine Operations) Group was formed on 1 April 1918 with its HQ at Haverfordwest. It had oversight of Llangefni, Killeagh, Malahide and Pembroke (airship stations), Fishguard (seaplane base) and units at Bangor/Aber, Dublin (RAF Tallaght), Milford Haven (HQ 77 Wing), Pembroke Dock and Wexford. As re-organisation spread northwards, Luce Bay (aeroplanes) and Luce Bay (airships) came firstly (from 1 July 1918) under the control of No. 22 (Marine Operations) Group with HQ at East Fortune, and then (from 12 August 1918) under a new 25 (Marine Operations) Group, HQ Luce Bay, which had oversight of nine Royal Naval Air Stations in south-west Scotland and Ulster.

Within 14 Group ground organisation started with six Coastal Patrol or Special Duties Flights (SDF) established on 6 June 1918, 519 and 520 SDFs at Pembroke, 521 and 522 SDFs at Llangefni and 523 and 524 SDFs at Luce Bay. On 25 July 1918 the six flights came under the umbrella of newly formed 255 Squadron, HQ Pembroke and the first CO Captain R. R. Soar DSC, a Western Front veteran who had flown seaplanes, Sopwith Camels and Triplane Scouts with 8 Squadron. He ran a Sopwith F1 Camel (D9542) as his own personal transport to get round his widely flung command.

In the first instance only eight DH6s operated out of Llangefni (B2791, B3020, B3021, C2021, C6656, C7861, C7863 and C7864). They were picketed out in the open, no hangars being provided. The SDFs as yet did not have their own armourer, nor mechanics, but were serviced by airship ground crews. The first fatality occurred on 14 August 1918 — the day before the unit transferred to its new home at Glan-y môr-isaf — when Airco DH6 B3021 suffered engine failure whilst on patrol and crashed, killing Lt. J. R. Johnstone. On the same day C2021 failed to return from patrol, lost at sea without trace. It was a rather sombre crowd that moved across the Menai Straits on the Thursday to their new war station.

The move coincided with, indeed, was prompted by, a further shake-up of Special Duties Flights and Squadrons. 255 Squadron, instead of being split between three widely dispersed stations and working under three Senior Naval Officers, was reconstituted as three squadrons, with the addition of an extra flight and more aircraft and pilots — if not the ground crews to keep them in the air! The official date for the transfer is 15 August 1918, although some aircraft may have flown in a day or two earlier. With a new coastal patrol flight, 530 SDF and six extra aircraft., they reformed as 244 Squadron under the command of Major H. Probyn. A much streamlined 255 Squadron now comprised just the two Pembroke SDFs. With the addition of 245 (Seaplane) Squadron based at Fishguard and formed on 20 August, the three new squadrons came under the aegis of No. 7 Operational Wing established on 8 August with HQ at Milford Haven. With the addition of a new coastal patrol flight (No. 529 SDF) 523 and 524 SDFs at Luce Bay formed a new 258 Squadron, but under the control of No. 25 (Marine Operational) Group. There was no Operational Wing within this latter Group.

After only two months at Aber 530 SDF, or at least a section of it, was sent to Ireland to patrol the eastern seaboard north and south of Dublin, the equivalent of the Malahide detachment of Llangefni's airships. On 18 October 1918 five DH6s C7786, C7796, C7861, C7863 and C9444 — flew out together under their own steam, a not inconsiderable achievement for obsolete aircraft! They were based at RAF Tallaght, itself only opened in August 1918 as a Training Depôt Station but in 1919 successively home to 117, 141 and 149 Squadrons in their internal security role.

Whilst at Tallaght 530 SDF lost C9444, piloted by Ft. Lt. H. F. Monypeny, which crashed near Dún Laoghaire on 6 November. The Armistice officially removed the need for special coastal patrols. Without realising it, the end of the war was already in sight, all U-boats having been recalled to Germany on 21 October. There was some 'tidying up' to do, but airships could seek out and destroy floating mines far more effectively and on 11 November 530 SDF's Irish detachment flew back to Aber, there to sit out two months of utter boredom in appalling conditions as the weather broke and tented accommodation became untenable, necessitating a quick move to outbuildings on the Penrhyn Castle estate.

Back on the mainland 521 SDF also suffered losses and sundry mishaps. The DH6 serial B3023 seemed particularly fated. On Monday, 26 August 1918, it crashed shortly after take-off on a rigging test (pilot Lt. A. C. Bencher with Air Mechanic Nichols in the observer's seat). Three weeks later, on Wednesday, 18 September, it stalled on a turn as it was taking off and spun in from 200 ft.. Capt. D. A. Tuck was slightly injured, but Air Mechanic Nichols later died of his injuries in Bangor Hospital. On the same day C6655, flying as a single seater, suffered engine failure and force landed in the sea some eight miles north-west of Holyhead. Luckily the DH6 had good

No. 14 (Marine Operational) Group
(formed 1.4.18, HQ Haverfordwest)

No. 77 Wing
(formed 8. 8. 18, HQ Milford Haven)

255 Sqn.	519 SDF ('A' Flt.) Pembroke / 520 SDF ('B' Flt.)	>	519 SDF ('A' Flt) / 520 SDF ('B' Flt.) — 255 Sqn. Pembroke
	521 SDF ('A' Flt.) Llangefni / 522 SDF ('B' Flt.)	>	521 SDF ('A' Flt.) / 522 SDF ('B' Flt.) / 530 SDF ('C' Flt.) — 244 Sqn. Bangor/Tallaght
	523 SDF ('A' Flt.) Luce Bay / 524 SDF ('B' Flt.)		426 Seaplane Flt. / 427 Seaplane Flt. — 245 Sqn. Fishguard

14 Group Coastal Patrol Organisation, April 1918–March 1919

A much patched SSZ.50 was possibly the hardest working 'Zero' at Llangefni, clocking up 775hrs. between 13 March 1918 – 24 January 1919.

floating characteristics and the pilot was taken off by an armed launch out of Holyhead. Sadly the aircraft eventually sank whilst under tow. On Friday, 11 October 1918, C2074, piloted by Lt. D. J. Wilks was lost over Caernarfon Bay. More lucky was Lt. C. W. S. Hall of 522 SDF, whose aircraft (F3350) crashed into the sea whilst patrolling a safe lane on 23 August 1918 — underneath the nose of a Holyhead destroyer which picked him up none the worse for wear.

Such tangible losses can be entered on the debit side of a squadron's books. On the credit side there is little in the way of positive results to show for four months hard slog. It was little comfort to 244 Squadron crews in their clapped out DH6s tucked away on the Menai Straits that the Short 184 seaplanes of 245 Squadron operating out of Fishguard could attack and claim, rather optimistically as it turned out, the destruction of at least four enemy submarines, three submerged and one on the surface which offered a spirited resistance before diving. This is not surprising since the wolf packs were queuing up in St. George's Channel, the doorway to the Irish Sea. But there is no doubt that the 'scarecrow' coastal patrols, however monotonous and unglamorous, achieved their purpose. In April 1918, before the SDFs were formed, there had been ten ships sunk in the Irish Sea. In October, admittedly when U-boat activity was declining, there were only two, the final fling before the collapse of Germany and both on days when the weather was such that neither airship nor aeroplane could operate. Overall it was a job well done, all the more creditable in view of hasty organisation, deficiencies of aircraft., crews and equipment.

Parallel to the introduction of aircraft into the coastal patrol scenario, were the continued attempts to improve the operational capability of the non-rigid airships. As noted above, there had only been a limited production of the 'SSP' class, of which Llangefni had received half. As the U-boat war reached its zenith so appeared in June 1917 the first of the 'Zero' ('SSZ') class. Their dimensions and capacity were roughly the same as those of the 'SSPs', but they were now powered by the 75 h.p. Rolls Royce 'Hawk' engine (supposedly trouble free!), and the car was purpose designed — water-tight, boat-shaped, strengthened keel doing away with skids or 'bump bags' (air cushions), aluminium on ash frame, and with three cockpit openings for W/T operator/air gunner forward, pilot centre and engineer aft with his bomb racks.

It was, however, sixteen months before the first 'SSZ' found its way to Llangefni and the 'SS' and 'SSP' types started to be written off or retired to training stations. Priority had been given to war stations at Polegate, Folkstone (Capel), Mullion and Pembroke responsible for escorting Channel convoys and transports to Boulogne. First to arrive on 13 March 1918 were SSZ.50 and SSZ.51, both of which would spend much of their time at Malahide mooring-out station. After an

inauspicious start surviving a torn envelope (by propeller) on 2 April and a forced landing on 5 April, SSZ.51 would clock a massive 655 flying hours in just 186 days before being lost at sea on Sunday, 15 September. Fortunately the three crew were picked up by USS *Downs*. It was replaced on 12 November by SSZ.72 which had just completed its trials at Howden. By the end of March the number of 'Zeros' on strength totalled four, with three more arriving in August.

SSZ.73 had the doubtful distinction of arriving in 'kit form' and being assembled at Llangefni and trialled there on 7 September 1918. Having flown the long way round — via Cranwell, Howden and Barrow — SSZ.72 arrived the day after the Armistice and would, along with SSZ.34 be deflated for good on the 26 November. Nevertheless, in the fourteen days she was operational SSZ.72 logged some 55 flying hours. Malahide, now that U-boats had disappeared from the Irish Sea, had closed on Friday, 22 November. The remaining four 'Zeros' on strength (SSZ.31, 33, 50 and 73) were retained for mine searching, atrocious weather permitting! They were finally deflated on 21 January 1919, two days after 244 Squadron (or its remnants) at Aber was also officially disbanded.

In her six months at Llangefni SSZ.34 flew $859^{1}/2$ hours, most of them cold, monotonous uneventful hours, except for a bit of excitement on Monday, 27 May 1918 when three of Llangefni's blimps and two destroyers, trying not to make too much smoke, were quartering Caernarfon Bay after a submarine had bee sighted some five miles off Pontllyfni. SSZ.34 dropped her 230lb bomb on rising bubbles of oil now, if it were the same U-boat, some 15 miles north-west of Nefyn, but with inconclusive results. Possibly it may have escaped into the open sea where two hour later, SSZ.51, working with a destroyer also depth bombed some suspicious oil bubbles, again inconclusively. In all likelihood the three sightings may have been one and the same U-boat lingering doggedly on the fringes of the safety lanes.

There was by now a subdued air of excitement and anticipation about Llangefni airship station and a new keeness to fly. Their first considerable contact with a U-boat had been made nine days earlier, on Saturday, 18 May when SSZ.50 had located a submarine off the Skerries, betrayed by an oil slick on the surface. The blimp had risen 300 ft. for better effect and dropped two 100lb and 230lb bombs. Messages were flashed to other airships in the vicinity and to a patrolling destroyer. The latter, *DO-1*, patterned the area with depth charges, possibly inflicting some damage but insufficient to force it to the surface.

At first light the following day, Whit Sunday, the search was resumed by three Llangefni airships. Twelve miles south-west of

LLANGEFNI NAVAL BLIMPS, 1915–18

These were non-rigid airships of various types, numbered separately with appropriate prefix letters according to type. Although small, simple and rather primitive, they were effective deterrents. All were scrapped during 1919.

Sea Scout/Submarine Scout Type (SS)

These were relatively small, makeshift airships, 145ft long, 27ft beam, >60,000 cu. ft. Cruising speed 40 m.p.h., endurance 8 hrs. at half throttle, two crew — pilot, WO/AG. Cost £2,550 Distinguished by different types of aeroplane fuselage (complete with wheels and skids) slung beneath the gasbag as car or nascelle. Radio, bombs, and larger petrol tank added.

BE. 2c type fuselage car with 70/75 h.p. Renault engine

SS.18   On strength 26.9.15; lost at sea 22.10.16 during landing attempt.
SS.22   Original establishment 5.11.15; to Wormwood Scrubs, -. 4.17.
SS.24   Original establishment -.-.15; to Luce Bay 11.7.17; deleted 17.7.18.
SS.25   Original establishment -. -.15; deleted Llangefni 15.3.18.

Maurice Farman type fuselage car with 75 h.p. Rolls Royce 'Hawk' engine

SS.33   From Luce Bay 4.11.16 replacing SS.18; to Wormwood Scrubs early 1918.

Sea Scout Pusher/Submarine Scout Patrol (SSP)

Modified SS types with Maurice Farman fuselage type car; three crew — pilot, engineer, WO/AG; >70,000 cu. ft. capacity; initially fitted with rear mounted 100 h.p. Green engines. Of the six built three were sent to Llangefni. Only these three survived the war.

SSP.1   On strength 5.7.17; to Cranwell 26.5.18 for training purposes.
SSP.5   On strength 18.1.17; as training craft to Cranwell, 26.3.18.
SSP.6   On strength 16.6.17; experimental to Pulham (Norfolk), 12.4.18.

Sea Scout Zero/Submarine Scout Zero Type (SSZ)

The Zero Type was the new standard version of the patrol blimp put into large-scale production (77 built, construction of a further sixteen cancelled) early in 1918. Rear mounted 75 h.p. Rolls Royce 'Hawk' as standard. Crew of three in specially designed car that could float. Endurance: 16hrs/50 m.p.h. > 40hrs/20 m.p.h..

SSZ.31   From Howden (Yorks.) 1.8.18; deflated 24.1.19.
SSZ.33   From Howden 1.8.18; deflated -.1.19.
SSZ.34   Original establishment 21.3.18; deflated 26.11.18.
SSZ.35   Original establishment 23.3.18; lost at sea 17.10.18; deleted.
SSZ.50   Original establishment 13.3.18, including Malahide (Dublin); deflated 24.1.19.
SSZ.51   Original establishment 13.3.18 including Malahide; lost at sea 15.9.18; deleted.
SSZ.72   Original establishment 12.11.18 including Malahide; deflated 26.11.18.
SSZ.73   Original establishment, assembled in situ 27.8.18 –7.9.18; deflated 24.1.19.

REPRESENTATIVE AIRCRAFT 244 & 255 SQUADRONS, 1917–18

A7654	Airco DH4.	Ferried in to Llangefni 7.11.17. Crashed on landing. Pilot killed.
B2791	Airco DH6.	On strength 244 Sqn. 15.8.18.
B2937	Airco DH6.	On strength 244 Sqn. 14.11.18
B2971	Airco DH6.	244 Sqn. Engine failure, forced landing 7.9.18.
B2973	Airco DH6.	On strength 244 Sqn. -.10.18.
B2976	Airco DH6.	On strength 244 Sqn. -.9.18.
B2977	Airco DH6.	On strength 244 Sqn. 31.8.18.
B2978	Airco DH6.	On strength 244 Sqn. 29.10.18.
B2979	Airco DH6.	On strength 244 Sqn. 1.11.18.
B2982	Airco DH6.	On strength 244 Sqn. 6.11.18.
B2983	Airco DH6.	On strength 244 Sqn. 23.11.18 (from 256 Sqn. Elswick).
B3010	Airco DH6.	On strength 244 Sqn. -.-.18 (from Nº 2 Wireless School, Penshurst).
B3020	Airco DH6.	On strength 255 Sqn. 14.8.18.
B3021	Airco DH6.	255 Sqn. Crashed on return from patrol 14.8.18
B3023	Airco DH6.	244 Sqn. Crashed 26.8.18; spun in 18.9.18; Air Mech. killed.
B3025	Airco DH6.	On strength 244 Sqn. 29.10.18, 8.12.18.
C2021	Airco DH6.	255 Sqn. Lost at sea 14.8.18.
C2074	Airco DH6.	244 Sqn. by 12.9.18. Crashed in Caernarfon Bay 11.10.18.
C6560	Airco DH6.	On strength 244 Sqn. 2.9.18, 10.12.18.
C6655	Airco DH6.	On strength 244 Sqn. 29. 8. 18; E/F, enforced ditching 18.9.18.
C6656	Airco DH6.	On strength 244 Sqn. 15.8.18
C7786	Airco DH6.	On strength 530 SDF, 244 Sqn. Tallaght, Dublin, -.10.18
C7796	Airco DH6.	On strength 530 SDF 244 Sqn. -.10.18.
C7800	Airco DH6.	On strength 530 SDF 244 Sqn. 2.9.18.
C7861	Airco DH6.	On strength 522 SDF 255 Sqn. 7.8.18; 244 Sqn., Tallaght, 18.10.18
C7862	Airco DH6.	On strength 244 Sqn, 19.8.18.
C7863	Airco DH6	On strength 255 Sqn. 9.6.18; 530 SDF 244 Sqn., Tallaght, 18.10.18.
C7864	Airco DH6	On strength 255 Sqn. 9.8.18 (from 131 Sqn., Shawbury).
C9444	Airco DH6	On strength 530 SDF 244 Sqn, Tallaght, 24.10.18; crashed Dalkie, Dún Laoghaire, 6.11.18
C9447	Airco DH6.	On strength 244 Sqn. -.11.18, 8.12.18.
F3350	Airco DH6.	244 Sqn. Crashed in sea, 23.8.18.

LLANGEFNI BLIMPS ON STRENGTH

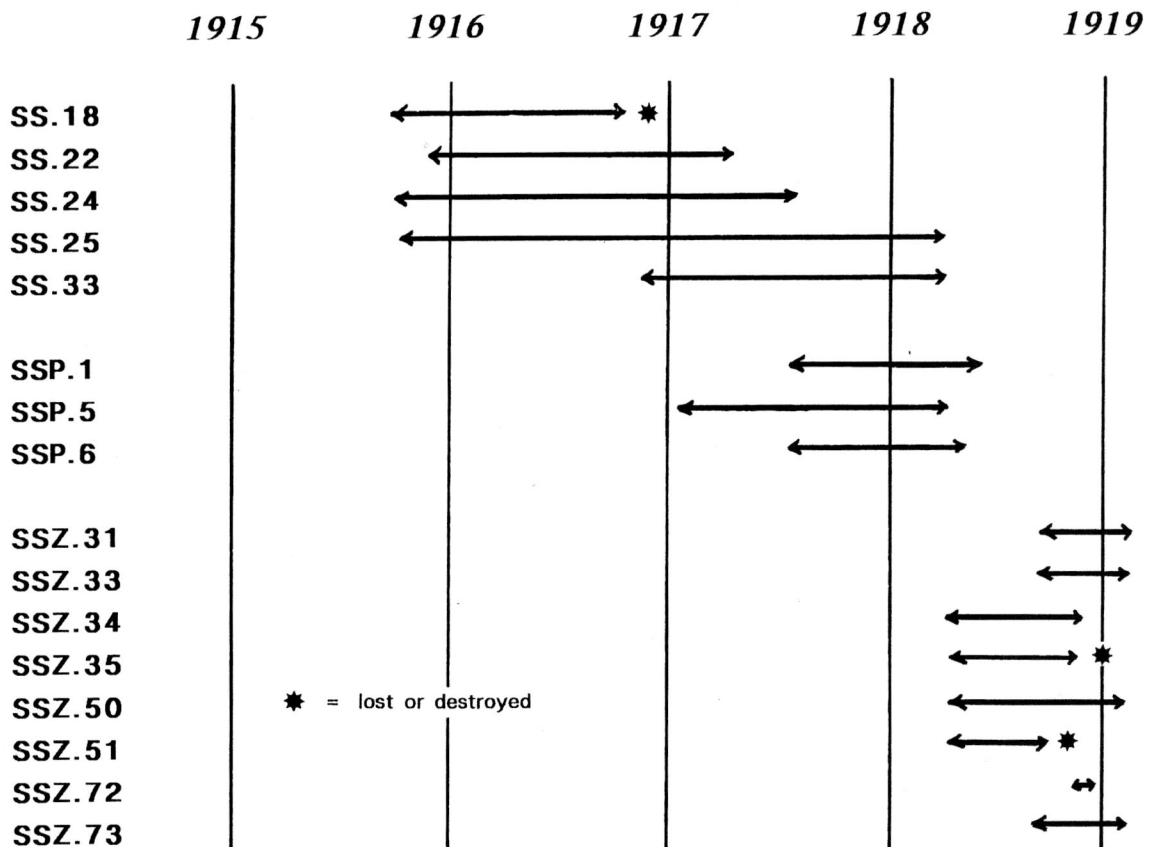

	1915	1916	1917	1918	1919

SS.18
SS.22
SS.24
SS.25
SS.33

SSP.1
SSP.5
SSP.6

SSZ.31
SSZ.33
SSZ.34
SSZ.35
SSZ.50
SSZ.51
SSZ.72
SSZ.73

✷ = lost or destroyed

Bardsey SSZ.51 picked up the scent again — the tell-tale wake of a periscope and the dark shadow of a submarine travelling slowly southwest just below the surface. A 230lb bomb was dropped, but to no avail. SSZ.35 added her bomb load, again to little effect. In a case of extreme overkill it was left to three American (Nos. *20, 40* and *66*) and one British destroyer (*DO-1*) to plaster the area with depth charges, bringing oil, air bubbles and debris to the surface — a sad end for UB-*119* which, under the command of *Oberleutnant zur Zee* W. Kolbe and with a crew of thirty-four, had sailed from Heligoland on Saturday, 27 April, for the Southwest Approaches and the Irish Sea killing area and so it proved! There were no survivors. Her last reported position to German Naval HQ had been on 25 May when she had sunk a steamer off Belfast. UB-*119* had been launched on 13 December 1917 and commissioned on 9 February 1918. It was standard practice to plaster a sunken U-boat with depth charges in order to break it open and release debris to the surface as confirmation of actual destruction and to possibly discover its identity.

This was a classic illustration of the role played by the 'scarecrow' blimps — deter, drive below the surface, observe, mark and call up reinforcements. For an airship to attack a submarine on the surface was in practice asking for trouble — blimps offered huge, irresistible targets to a U-boats gun crew. Sometimes, after hours of stultifying boredom, caution would be thrown to the wind — as happened on 17 December 1917 when SSZ.16, based at Pembroke, attacked with her forward Lewis gun the U-boat that had dared fire upon her! The machine gun was not even a defensive weapon, just a useful device

worked by the W/T operator to detonate floating mines! To use it to attack a submarine... The mind boggles!

The events of these three days, 18, 19 and 27 May, were the apogee of Llangefni's anti-submarine patrols. Thereafter no further sightings, never mind bombings. For SSZ.35 routine tedium would be broken by even greater mind-bending monotony an attempt to set up an endurance record for a dirigible. On 28/29 June 1918, in ideal weather conditions and commanded by F/Lt. T. B. Williams with F/Lt. Farina as co-pilot, she slowly flew clockwise around Llangefni's patrol area — Carlingford Lough–Luce Bay–Isle of Man–Ribble estuary– Mersey Lightship–Holyhead–Cardigan Bay and home, hastened at the end by a little tomfoolery, a faulty fuel gauge and a leaking envelope. Nevertheless their record of 26 hrs. 10 mins. would stand until 11/13August when SSZ.39, based at Polegate in Sussex and captained by F/Lt. Bryan, flew 1,000 miles in 50 hrs. 55 mins. SSZ.35 would not see the end of the war. On 17 October 1918 with 822 flying hours on the clock she was lost at sea (crew saved) some fifteen miles northwest of Holyhead, her envelope torn (cause unknown), an end almost as ignominious as her introduction to a bewildered public five months earlier!

Perhaps a fitting round-off to combined operations, in the party-fuddled euphoria of the week following the Armistice, was the flying of SSZ.73 underneath the Menai suspension bridge, Major Thomas Elmhirst, CO Llangefni, at the helm, Captain Gordon Campbell, Senior Naval Officer, as observer, and a very worried Air Mechanic Charles Jones in the engineer's cockpit, all the more worried since he was a local man and knew the Menai Straits, its rip tides and its famous bridges. It was one thing for Major Probyn, CO 244 Squadron to take his senior officer under the bridge in a DH6, but a 'Zero' airship... ?

But as Air Marshal Sir Thomas Elmhirst KBE, CB, AFC later wrote in his *Recollections*, it was a simple matter of reconnaissance, measuring the height from bridge deck to surface of water at low tide, 150ft or thereabouts. The height of the airship was known, 30ft for envelope, add 15ft for car and rigging. Dangle over cockpit side a small sandbag on a length of cord measured to allow safe passage of blimp underneath the bridge, steer for centre of said bridge with sandbag skipping surface of water and 'Bob's your uncle!' Unlike the pilot of a DH6, Major Elmhirst would not have been able to see the bridge deck because of the envelop over his head.

Llangefni's U-boat menace

As noted earlier, the Naval Airship Patrol Stations at Llangefni, Luce Bay and Pembroke were established to combat the growing German submarine menace in their respective quadrants of the Irish Sea. A brief consideration of the nature of this threat as it manifested itself in Llangefni's and 244 Squadron's patrol area will therefore not come amiss at this point.

In the First World War some 178 enemy submarines were sunk (compared with 784 in the Second World War) with the loss of over 5,400 officers and men (compared with nearly 27,500 in the later conflict). Relating to the 1939–45 struggle the Navy Staff Historian would reluctantly conclude that 'the Allies never gained a firm and final victory over the U-boats'. Even as Russian forces readied themselves for the final onslaught on Berlin the *Kriegsmarine* launched a new submarine offensive from its Norwegian bases. Twenty-five U-boats in British waters pinned down some 400 ships and 800 aircraft.

Things were slightly better in 1918. If not mastered, the U-boats were controlled up to a point. The political configuration of belligerence eventually made for a tighter rein on the movement of enemy submarines by means of 'gates' or 'choke points' *e.g.* the Straits of Dover, the Shetlands-Norway gap — which U-boats had to negotiate to reach their patrol areas. But while such portals were secured by standing surface patrols, extensive minefields, nets, sweeps etc., enemy submarines with daring and able commanders still managed to get through into the Irish Sea to attack coastal shipping and stragglers from dispersing eastward-bound Atlantic convoys.

The accompanying map of British merchant ship losses, with the approximate limits of Llangefni's patrol area highlighted, shows the Irish Sea to have been a happy hunting ground for German U-boats. An obvious pattern is immediately discernible. The main Liverpool trade artery through St. George's Channel is cluttered with wrecks with major nodes off Bardsey Island, in Caernarfon Bay off South Stack, and along the north coast of Anglesey, especially off Point Lynas. The treacherous tides and shifting banks off the Mersey and Dee estuaries seem to have been sensibly avoided by enemy submarines. At a later date Irish Sea traffic would be filtered at strategic points into a rudimentary one-way system, but even in 1915 a subsidiary route along the east coast of Ireland was easing the congestion somewhat. Enemy submarines lurked off Waterford, Wexford, Arklow and Dublin to obvious good effect, preying both on the east-west coastal trade and ferries, and vessels that had shed their escorts as convoys broke up for the various western seaports.

A provisional complementary list of shipping losses, abstracted from *Lloyd's Register*, shows some seventy-six merchant ships and/or naval auxiliaries sent to the bottom in Llangefni's patrol area alone, six in 1915, three in 1916, twenty-six in 1917, and thirty-eight in the ten months of 1918 before recall to Germany. In most cases the identity of

offending submarines is known. They were not always the more modern '*U*'–class. Some of the most successful were the small and obsolescent vessels, the '*UB*'s, fit only for coastal work, and the '*UC*'s or minelayers.

On the other hand U-boat losses in the Irish Sea itself were few, *U87* (sunk 25 December 1917) and *UB119*, already noted as being sunk in a combined operation off Bardsey Island on 19 May 1918. Three were sunk in the North Channel — *UB82* (17 April 1918), *UB85* (30 April 1918) and *UB124* (20 July 1918) — and a further four — *U58* (17 November 1917), *U61* (26 March 1918), *U106* (7 October 1917) and *UC33* (27 September 1917) — in St. George's Channel between Milford Haven and Wexford, the latter area and the Western Approaches being a favourite operating area for U-boat 'group' operations. Possibly the term 'wolf pack' is best reserved for the Second World War. One reason for the discrepancy between these two sets of statistics, apart from the obvious advantages of invisibility and surprise enjoyed by submarines, is that allowances have to be made not only for the proper maintenance of U-boats and respite and rest for crews, but also for the time consuming voyages to the hunting grounds off the western seaports. Only two or three U-boats would be operating at any one time in the wide expanse of the Irish Sea. One can almost work out the rotas from the U-boat numbers given in the attached list! Bearing in mind that the latter is artificially contrived to embrace roughly Llangefni's later operational area, some thirty-five different U-boats are noted. Twenty-three of these would survive the war. The remainder would be lost on other patrols or, by cruel stroke of fate, on the homeward voyage from the Irish Sea.

Thus *UB123*, commanded by *Oberleutnant zur See* R. Ramm, having sunk the Irish Mail ship *Leinster* on 10 October 1918 and the SS *Dundalk* off Anglesey on 14 October, turned for home. These losses had occurred when bad weather had grounded both Llangefni's blimps and 244 Squadron's D.H.6s. *UB123* successfully negotiated the North Channel, avoiding Luce Bay's Zeros and Belfast's surface patrols. On 17 October it had rounded Scotland and exchanged signals with *UB125* about a safe route through the newly laid Northern Barrage. This was the last seen or heard of the submarine. The boat and her thirty-six crew were probably lost to a mine, possibly not diving deep enough! The tragedy was that *UB123* was, one of the last three U-boats to be sunk in the First World War — two others were sunk on 28 October. With luck, her crew could have made lt back to base, just in time for the surrender and internment of the German High Seas Fleet on 21 November 1918.

Similarly, a week after sinking the SS *Tarbetness* in Caernarfon Bay on 7 March 1918, the *U110*, commanded by *Korvettenkapitan* K. Kroll, was at the bottom of the Atlantic off Malin Head. Kroll had dallied rather too long on the northern trade and convoy route into the Irish Sea. The sinking of the SS *Amazon* on 15 March alerted two Royal Navy destroyers who carried out a successful depth charge attack. Badly damaged, *U110* surfaced and was abandoned, but only four of the thirty-nine crew survived their immersion in the Atlantic.

Such ignominious ends notwithstanding, the rewards were well worth the risk and within this land-girt sea a daring U-boat captain could literally have a field day. From the earliest months of the war when enemy submarines were seeking purely naval targets, commanders had frequently commented in their reports on the volume of merchant shipping entering the Mersey, Clyde and Bristol Channel. One notes, for example, *U103* having two or three bites at the cherry. Commissioned in July 1917, she was responsible for the sinking of the SS *Cork* off Point Lynas, Anglesey, on 26 January 1918 with the loss of twelve lives, and SS *Ethelinda* off the Skerries three days later, when twenty-six crew perished. She was back in the Irish Sea in March, sinking three more merchant ships along the Anglesey-Wicklow trade route, the last being SS *Kassanga* on 30 March 1918. Her next patrol, however, was not so fortunate. On 12 May *U103* was one of five U-boats lurking in a pack off the Lizard awaiting the arrival of an eastward bound convoy. Recently American troopships were being convoyed only as far as Fastnet Rock, proceeding thereafter independently, possibly relying on superior speed to dodge enemy submarines. Either the conning tower look-outs were not very vigilant or the darkness was very intense, but *U103*, lying on the surface, was run down by the White Star troopship *Olympic*, making a good 25 knots towards Southampton. *Kapitanleutnant* C. Rucker and thirty-four crew were picked up by the USS *Davis* and landed at Queenstown. Ten crew went down with their ship.

As one peruses the list of shipping losses one senses that here, perhaps, are several U-boat 'aces' in the making as a direct result of a finely judged and well-behaved torpedo. But to a layman some of the entries are rather puzzling. Three losses during the last two months of 1916 are credited not to capture, gunfire or torpedo but to mining, more specifically to mines laid by *U80* in the quadrant between Anglesey and the Isle of Man. How many mines on how many missions would have to be sown by *U80* in such a vast expanse of sea to guarantee at least one sinking, never mind three? Or were the Irish Sea trade routes so densely packed with merchant shipping that the chances on even a single mine claiming a victim were very high? Exactly how do official records associate a mine from a particular U-boat with the loss of a particular ship? Were locations of mines laid and ships lost plotted with such exactitude that post war researchers in the naval archives of two countries were able to pair them off? These are niggling questions beyond the ability of simple local historians to answer satisfactorily.

As far as our particular part of the world is concerned submarine warfare started on 30 January 1915 when, fifteen miles off the Morcambe Bay lightship, the SS *Ben Cruachan* (3,092 tons) was apprehended by *U21* en-route from Cardiff to Scapa Flow with a cargo of coal. The crew were permitted to lower and board their boats. Sea cocks were opened and, assisted by an explosive charge, the *Ben Cruachan* at once went down to the bottom. Owned by the Morrison Shipping Company of Newcastle she was only twelve years old. Even though the ship's flag was taken as a war-trophy, it was all very correct, carried out in accordance with international law and the traditional code of conduct at sea and, more importantly, at minimal cost. Canny U-boat commanders did not like wasting torpedoes. One notices *UC65*'s successful sweep of the Liverpool Bar–Bardsey–Arklow triangle in February/March 1917. Despite the vigilance of Llangefni's blimps *UC65* sent nine ships (16,408 tons) to the bottom, only three by torpedo. The rest were 'visit and search' incidents ending in scuttling hastened by gunfire or a well-placed bomb. It is interesting to ponder whether *UC65* had run out of torpedoes or its commander, *Kapitanleutnant* Klaus Lafrenz, even at this late stage in the war, was still imbued with some of the ancient chivalries.

But on 4 February 1915 a more ruthless dimension had been given to the war at sea. Following the disappointing performance of their surface raiders at large, Germany decided to prosecute an unrestricted campaign against allied merchant shipping. A war zone was declared around the British Isles within which ships including neutrals, were liable to be sunk without warning. In one sense it was a logical step. For a submarine to operate on the surface and to give fair warning of intention like any other surface vessel, was to cancel out a U-boat's one real advantage, the very reason for its existence, its invisibility and stealth and surprise in attack. On 20 February SS *Cambank* (ex-Raithmoor 3,112 tons), Liverpool bound with a cargo of copper ore, was torpedoed and sunk by *U30* ten miles east of Point Lynas, Anglesey, the first victim in our area of the new 'get tough' policy.

Over the next three months seven more ships would be sunk in the Irish Sea and St. George's Channel, a creditable total considering that in the first half of 1915 Germany could only boast some twenty operational submarines. However, by the end of December 1915 a further sixty-one new craft, would have been added to the submarine

fleet. But in the wake of the international outcry and clamour following the sinking of the *Lusitania* on 7 May and the *Arabic* on 19 August, both with the loss of American lives, Germany had perforce to modify its U-boat campaign again, temporarily prohibiting attacks on passenger ships of any nationality without giving prior warning and ensuring the safety of passengers.

There would only be nine further sinkings in the Irish Sea and approaches in the three months June –August 1915 and all followed the 'visit and search' procedures deemed politic under the new policy. One notes, for example, how prudent was the *U38* as it swept up the Irish Sea sending five ships to the bottom on 17 August 1915 — SS *Queen* and SS *Bonny* in St. George's Channel, SS *Glenby* off the Smalls, all three captured and sunk by gunfire, and SS *Kirkby* and SS *Paros*, captured, searched, abandoned and then sunk by well placed torpedoes.

Following *U38*'s exploits there were no sinkings in the Irish Sea for the rest of the year. Fearful of offending further American opinion and possibly bringing her into the war on the side of the Allies, Germany withdrew all its U-boats from the Irish Sea and western waters rather than its submarine commanders be hamstrung by restrictions. This 'lull' coincided with the first four months of NAPS Llangefni's existence, permitting an extensive 'shaking down' period as the station took delivery of its first four Sea Scout blimps. However, U-boat inactivity would continue throughout the greater part of 1916. Not being privy to the mutually destructive conflicts between the German Chancellor, Foreign Minister, Kaiser and the 'hawks' of the Naval High Command, Llangefni's airship crews ploughed on, routinely dissecting their monotonous quadrant of the Irish Sea to no avail. Notwithstanding bombing and machine gun practices and other diversions such as the loss of SS.18 and the advent of SS.33, this led inevitably to frustration and some loss of morale, and goes far to explain the heavy drinking and boisterous behaviour of officers and ratings off-duty in the tiny Welsh, essentially Nonconformist and abstinent, market town of Llangefni. Police and magistrates were hard pressed at times, and local hack reporters never lacked for fresh copy.

Unknown to the Admiralty, diplomatic rather than military and strategic pressures weighed paramount in the prosecution of the U-boat war. The German Naval High Command, supremely confident in the efficacy of its rapidly expanding submarine flotillas, was pressing for the re-introduction of an unrestricted economic blockade by U-boats, Admiral Henning von Holtzendorff, German Chief of Naval Staff, predicting the breaking of British resistance within six months at the outside. The 'hawks' won the battle for the Kaiser's ear. A compromise was reached and a new U-boat campaign authorised on 11 February 1916 — but again with restrictions and therefore again doomed to failure from the start.

Only shipping in the war zone was to be attacked without restriction. Outside this zone merchantmen could be sunk without warning if they were armed; passenger liners were not to be touched. Four enemy submarines were in the Irish Sea and approaches in February/March 1916 but were slow to open their tallies. On 27 March the SS *Manchester Engineer* (4,302 tons) was torpedoed and sunk in St. George's Channel by *U44* whilst on the last leg of its voyage from Philadelphia to Manchester. On 23 March the SS *Englishman* (ex-*Sandusky, Montezuma, Iona* 5,257 tons) making for Portland (Maine) out of Avonmouth, had been sent to the bottom by *U43* just as she had cleared the North Channel. Then, before havoc could be wrought into Llangefni's patrol area, a halt was called to the attacks. There would be no more sinkings in the Irish Sea until the fortuitous minings of December already noted.

The reason for this shortest-lived of U-boat campaigns was again political. On 24 March 1916 the French Dieppe–Folkstone passenger ferry had been torpedoed by the *UB29*. Although remaining afloat a number of Americans had been killed or injured. In its official protest Washington threatened an immediate break in diplomatic relations unless Germany abandoned its current method of submarine warfare against passenger ships and merchantmen or else. . . ! An alarmed German Government reacted swiftly, prohibiting all U-boat attacks unless observing the code of conduct — stopping a ship, examining papers and cargo, ensuring the safety of crew etc. — implicit in prize law regulations. Angry and exasperated, submarine flotilla commanders withdrew all their operational U-boats from western waters. Enemy submarines would be back in the Irish Sea with a vengeance in 1917, beginning with the *U55* and *UC65* in January/February. But for the moment merchant ships moved freely in and out of Liverpool and the west coast ports and counter measures were correspondingly relaxed. On 18 January 1917, just in time to meet the resurgence in U-boat activity, Llangefni would take on strength SSP.5, the first of its three stop-gap 'Sea Scout Pushers' that would hold the fort until the arrival of the SSZs in March 1918.

These early, half-hearted attempts at submarine warfare, hampered and abruptly curtailed as they were, served notice on the Admiralty to prepare for future eventualities. By the end of 1916 it was becoming evident that an all-out U-boat war was Germany's only real option if it was to survive. The Battle of Jutland (31 May–1 June 1916) had been indecisive; German armies had failed to break through on the Western Front; there was near economic bankruptcy at home, and the British blockade was increasingly effective. Defeat from total exhaustion was imminent unless — the logic was brutally simple — Britain was broken first. And England's mainstay was her merchant shipping… !

On 1 February 1917 Germany, who now could muster some 154 up-to-date submarines, once more committed its flotillas to unrestricted U-boat warfare in western waters. By the end of that month British shipping losses had leapt to 86 compared with 35 (110,000 tons) the previous January. The casualty figure would rise to 103 (332,000 tons) in March and 155 during April, in which month some 516,000 tons of shipping went to the bottom via torpedo attacks, with a further 30,000 tons lost through striking mines. One can add to these catastrophic April figures another 323,000 tons of Allied and neutral ships carrying essential and war supplies. On top of ships thus taken permanently out of circulation, the economic dislocation due to ships being damaged, sailings postponed or cancelled, was acute. Masters were forced to make long diversions, that is, as and when they sailed at all. Ship movements in and out of British ports fell to under 300 a month compared with over 1,100 in the same period the previous year.

These were unacceptable losses, not sustainable for long. The *Kriegsmarine* was well on the way to achieving its avowed aims. But also during April two things occurred that would drastically cut these losses. Grinding down old prejudices and stifling dissident voices, the convoy principle was generally adopted. Then on 6 April America entered the war, ultimately providing sufficient destroyers and escort vessels to make the convoy system work. As the long forgotten art of marshalling and chaperoning vast armadas of heterogenous merchantmen were re-learnt and new lessons taken on board, shipping losses fell dramatically. Out of 800 ships convoyed in July/August 1917 only five were lost to U-boats.

Against these developments on the politico-economic front, the crews of Llangefni's blimps suddenly found life very hectic, with three ships sunk in its operational area in February, seven in March and twenty-eight by the end of the year with, frustratingly, no enemy submarine brought to book as compensation. The rot set in on Wednesday, 14 February 1917 and the next day when *UC65* sank the Dublin-registered SS *Ferga* (791 tons) en route from Swansea to Liverpool with a general cargo, the SS *Greenland* (1,753 tons) carrying Government stores from Fleetwood to Cherbourg, and the tiny SS *Kyanite* (564 tons) which, as its name suggests (= aluminium silicate), was outward bound from Fleetwood with a cargo of alkali. All three were sunk off Bardsey Island, right under the noses of Llangefni's

blimps. Bardsey was one of the points where Llangefni's and Pembroke's patrol area overlapped and there was as yet, if ever, no synchronisation of patrol lines and times between the two. But four blimps could not ride shotgun everywhere at the same time. Many masters, even in the era of convoys, elected to sail independently and take their chance relying on speed and their knowledge of Welsh coastal waters to outwit a lurking U-boat. Submarine commanders liked these 'loners'. *UC65* is also credited with laying the mines that sank the SS *Lycia* off St. David's Head on 11 February, the SS *Inishowen Head* in the St. George's Channel on 14 February, and SS *Afton* off Strumble Head on 15 February. *UC65* was back wreaking havoc in the Irish Sea on 24-28 March and 4 May 1917, but how many missions this represents is uncertain.

Apart from a couple of minings, shipping losses in Llangefni's quadrant fell markedly in the period April–October 1917, aided by some fine weather, intensified 'scarecrow' patrols, and increased flying time with the opening of the out-station at Malahide. In June and July the unit had lost SS.22 and SS.24 to other RNAS establishments and had taken delivery of SSP.1 and SSP.6. The implementation of the convoy system in the Irish Sea also began to pay dividends, making for a dearth of targets all round. However, this would backfire to some extent. U-boats, having little luck in the Western Approaches began to congregate in the Irish Sea where the irregular configuration of the coastline, that of Wales in particular, made the many bays self-contained hunting grounds where inshore shipping could be picked off at leisure. As noted earlier the Admiralty would attempt to combat this disturbing trend by the introduction in August 1918 of strategically located Special Duty Flights and/or Squadrons, but to little avail.

As bad weather set in and the 'scarecrow' blimps were grounded, shipping losses began to rise again, with five in October 1917, three of them off the east coast of Ireland as if to mock the Malahide arrangements. The period 28 November–28 December saw at least seven submarines at large in Llangefni's patrol area, working two or three at a time and notching up at least twelve sinkings between them.

The number of ships sunk and tonnages are but two indices of the success or otherwise of the U-boat campaign. A third index is the loss of life. Within Llangefni's quadrant in 1917 the recorded loss of life amounted to 270, rocketing to 551 in 1918. Loss of ships and crews were, of course subject to the usual strictures of wartime censorship, but along the North Wales coast small, close-knit maritime communities were fully aware of the dramas that were being played out just over the horizon off their stretch of coast. Bodies, some identifiable, others unrecognizable after long immersion, were washed ashore to receive Christian burial, a brief annotated entry in the parish register, and a nominal headstone in the local churchyard.

Cemeteries at Llanengan and Llangian on the eastern extremity of Llŷn, and at Llanaber, near Barmouth, each have graves for the victims from the sinking of SS *Memphian* and SS *Greldon*, both sent to the bottom by the *U96* on 8 October 1917. The Liverpool registered *Memphian* (6,305 tons) was Boston (USA) bound in ballast; thirty-two lives were lost. The Cardiff-based *Greldon* (ex-Dartmouth 3,322 tons) had just left Birkenhead with a cargo of coal for Italy. They had met up with the *U96* off the North Arklow lightship, Co. Wicklow. Roughly along this latitude tides meet and bodies cast into the sea are at the mercy of vacillating currents, in this particular case being carried some 65-85 miles before being dumped unceremoniously on a remote Welsh beach. Further up the Merionethshire coast at Llanddwywe five more bodies from the SS *Greldon* lie buried, Second Engineer H. Porteous and Fireman H. Williams; three other crewmen remain unidentified.

The long arm of the Llŷn peninsula stretched out to pluck other wretches from the sea. At Abererch, near Pwllheli, lies buried Fireman T. Williams from the SS *Poldown* which on 9 October 1917 struck a mine laid by *UC51* off Trevose Head, Cornwall. His body had drifted anything between 180–200 miles before coming to rest. Elsewhere on

MERCHANT SHIP LOSSES IN THE IRISH SEA 1915-18

North Channel

Belfast

Isle of Man

IRISH SEA

Dublin

Llangefni

Liverpool

Wexford

St. George's Channel

0   20   40

miles

• = sinkings

⌐ = Llangefni RNAS operational area

North Wales coast — Bangor, Colwyn Bay, and Abergele, — where they were buried. In other churchyards —Porthmadog, Llanfaelrhys, Prestatyn — lie the graves of victims of torpedoings and minings in the English Channel, but these have obviously been returned to their home towns and villages for burial.

The 1918 list of shipping losses in Llangefni's patrol area shows the same steady haemorrhaging, peaking during the winter's gales and storms which kept Anglesey's blimps grounded for days on end — 5 (4) in January 1918, 9 (1) in February, 7 (4) in March, 7 (4) in April. Figures in parenthesis indicate additional losses elsewhere in St. George's Channel and the Irish Sea. There was a marked drop in the number of casualties during the summer months, perhaps coinciding with the introduction of the new Sea Scout Zeros from March onwards and the arrival at Milford Haven and Holyhead of American light forces to ease the burden of providing patrols and convoy escorts. In June the arrival of the 'scarecrow' Special Duties Flights, first at Llangefni then moving to a temporary aerodrome outside Bangor in August 1918, may also have contributed to the seasonal fall in losses. It is difficult to quantify the achievement of these second rate machines with second grade aircrews. Their task was to patrol the off shore trade route between Anglesey and Liverpool Bar lightship, but the endurance of a DH6 was only some 90 minutes. At least one notices from the attendant list no ships were sunk along their patrol line! But possibly more pertinent to the fall away of losses were events on the continent where the last great German push on the Somme in March-April fizzled out through sheer exhaustion of men and supplies. At this moment Germany realised she was beaten.

On 14 September 1918 the *UB103* was the last enemy submarine to pass westwards through the Strait of Dover. Henceforward the Strait was effectively stitched up with U-boats forced to take the 'north about' route into the Irish Sea. Some managed it, but they were now few. Between August-October only five U-boats are noted in Llangefni's patrol area, but their commanders must have been brave, determined — if not desperate — men. Attacks are short, concentrated in time and space, as witness *UB92*'s depredations in Caernarfon Bay on 20-22 August or those of *UB64* making her last visit in September to a killing ground that had in happier times been so profitable.

On 21 October 1918 all German U-boats at sea were recalled. Some 176 submarines survived for the final surrender and internment in neutral or Allied ports. Two of the last three sinkings in the Irish Sea were in fact in Llangefni's 'home waters'. On 14 October the tiny *Dundalk* (794 tons), outward bound from Liverpool to Dundalk with general cargo and many deck passengers, was sent to the bottom by torpedoes from both the *UB123* and *U90* off the Skerries, Anglesey, with a heavy loss of life. *UB90* went on to sink SS *Pentwyn* off the Smalls, St. George's Channel, two days later.

On 18 October the Government impressed SS *Hunsdon* (2,899 tons), in ballast from Le Havre to Belfast, was sunk by *UB92* one mile off Strangford Lough lightbuoy. It is perhaps coincidence that, also on 18 October, 530 Special Duties Flight (otherwise 'C' Flight 244 Squadron) was detached from Bangor to RNAS Tallaght, Co. Dublin, to run 'scarecrow' anti-submarine patrols along the east coast of Ireland.

On 20 November 1918 a long file of German submarines

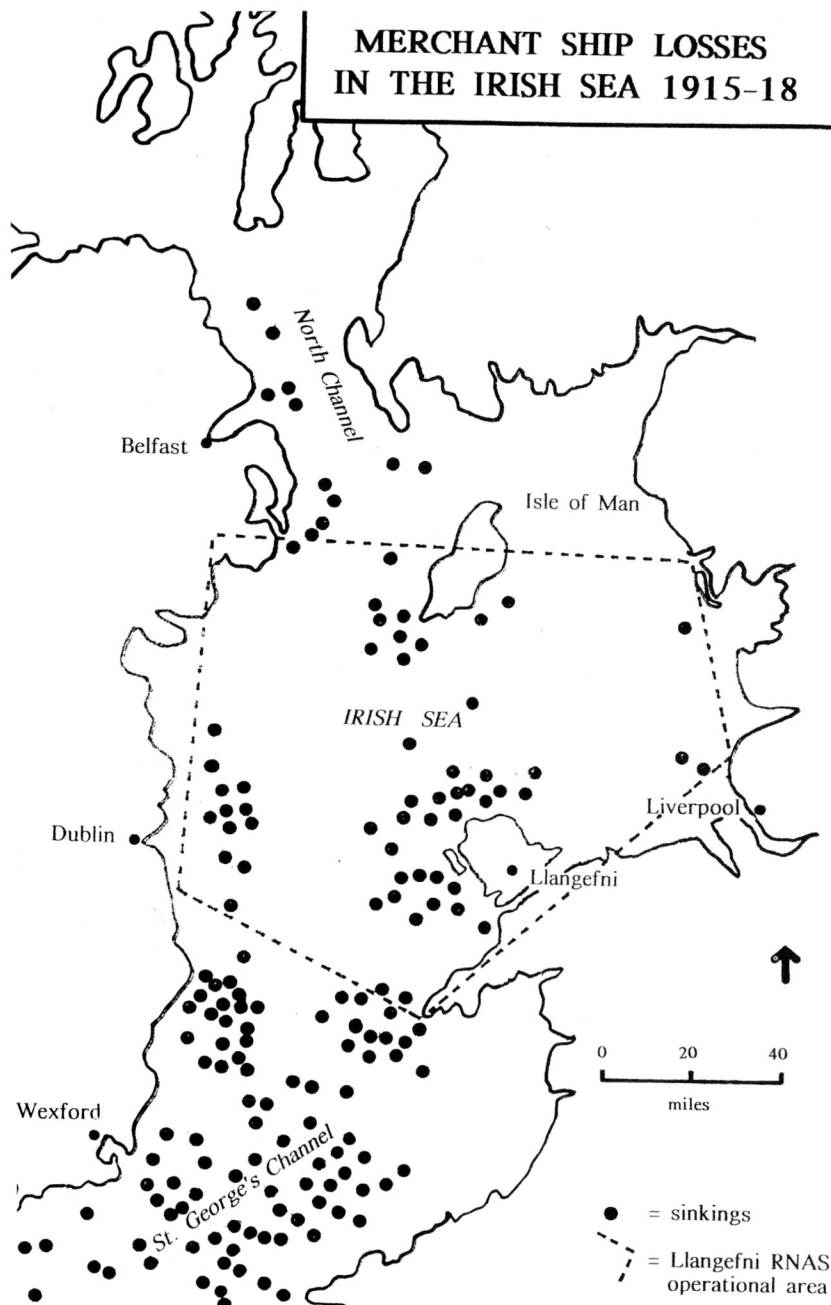

the peninsula Aberdaron and Edern cemeteries collected three victims, one unidentified, from HMS *Stephen Furness* (1,712 tons) torpedoed and sunk west of the Isle of Man by the *UB64* on 13 December 1917. Owned by the Tyne-Tees Steamship Company, Newcastle, she had been pressed into Government service as an armed boarding steamer. A fourth victim, AB W. F. Talmey, RN, lies buried in St. Mary's churchyard, Llanfair-yng-nghornwy, but his body was brought here after being picked up from the sea.

Tides and waves are unpredictable. Bodies from the Cork-registered SS *Serula* (1,388 tons) torpedoed and sunk by the *UB64* on 16 September 1918 off Fishguard lie buried at Aberffraw, Anglesey. The *Serula*, outward bound from Manchester to Rouen with general cargo, carried a crew of seventeen. Six crew members of the Irish ferry SS *Leinster* (176 lives lost) were buried at Holyhead where a rescue ship had landed them. With almost as great a loss of life was the African Steamship Co./Elder Dempster Line SS *Apapa* (7,832 tons) sunk off Lynas Point, Anglesey, by the *U96* on 28 November 1917. Longshore drift deposited bodies ashore at successive points along the

British Merchant Ships Sunk in NAP5 Llangefni's Patrol Area

1915

30.1.15 *BEN CRUACHAN* (3,092 tons). Captured and sunk by demolition charge by *U21* fifteen miles off Morcambe Bay lightship. From Cardiff to Scapa Flow with cargo of coal.

20.2.15 *CAMBANK* (3,112 tons). Torpedoed and sunk by *U30* ten miles E of Point Lynas, Anglesey (4 lost). From Huelva (Spain) to Garston Dock, Liverpool, with cargo of copper.

9.3.15 *PRINCESS VICTORIA* (1,108 tons). Torpedoed and sunk by *U20* sixteen miles N of Liverpool Bar lightship. Out of Aberdeen with general cargo.

13.3.15 *HARTDALE* (3,839 tons). Torpedoed and sunk by *U27* seven miles SE by E of South Rock (2 lost). Alexandria bound from Clyde with cargo of stone.

17.8.15 *PAROS* (3,596 tons). Captured and torpedoed by *U38* thirty miles west of Bardsey Is. Out of Karachi for Manchester with cargo of wheat.

17.8.15 *KIRKBY*(1,891 tons). Captured and sunk by torpedo by *U38* off Bardsey Is. Out of Barry with a cargo of coal.

1916

4.11.16 *SKERRIES* (4,278 tons). Struck mine laid by *U80* off the Skerries, Anglesey. Out of Barrow for Barry in ballast.

18.12.16 *OPAL* (599 tons). Struck mine laid by *U80* off Isle of Man (12 lost). Ship bound from Llanddulas to Clyde and Belfast with cargo of limestone.

19.12.16 *LIVERPOOL* (686 tons). Struck mine laid by *U80* off Chicken Rock (Isle of Man) (3 lost). Out of Liverpool for Sligo with general cargo.

1917

14.2.17 *FERGA* (791 tons). Captured and sunk by gunfire by *UC65* fifteen miles S of Bardsey Is. Out of Swansea for Liverpool with general cargo.

14.2.17 *GREENLAND* (1,753 tons). Captured and sunk by demolition charge by *UC65* off Bardsey Is. Sailed from Fleetwood for Cherbourg with cargo of Government stores.

15.2.17 *KYANITE* (564 tons). Captured and sunk by demolition charge by *UC65* off Bardsey Is. Making for Bristol out of Fleetwood with a cargo of alkali.

24.3.17 *FAIREARN* (592 tons). Captured and sunk by demolition charge by *UC65* 16 miles WNW of South Stack, Anglesey. Making for Garston Dock with a cargo of coal.

25.3.17 *ADENWEN* (3,798 tons). Torpedoed and sunk by *UC65* off North Arklow lightship (10 lost). Out of Cienfuegos (Cuba) for Queenstown and Liverpool with a cargo of sugar.

27.3.17 *KELVINHEAD* (3,063 tons). Struck a mine laid by *UC65* off Liverpool Bar lightship. Had sailed from Clyde via Liverpool for Buenos Aires with a general cargo.

28.3.17 *WYCHWOOD* (1,985 tons). Torpedoed and sunk by *UC65* off Wicklow coast, mouth of Avoca River. (3 lost). Bound for Scapa Flow with cargo of coal from Barry.

28.3.17 *SNOWDON RANGE* (4,662 tons). Torpedoed and sunk by *UC65* 25miles west of Bardsey Is. (4 lost). Out of Philadelphia for Liverpool with cargo of wheat and foodstuffs.

28.3.17 *ARDGLASS* (778 tons). Captured and sunk with demolition charge by *UC65* off South Arklow lightship. From Port Talbot for Belfast with a cargo of steel.

29.8.17 *LYNBURN* (587 tons). Struck mine laid by *UC75* off North Arklow lightship (8 lost). Bound for Whitehaven from Cork with a cargo of pit wood.

2.10.17 *LUGANO* (3,810 tons). Struck a mine laid by *U79* in North Channel off Antrim coast. Bound for Liverpool from Newport News (USA) with general cargo.

8.10.17 *MEMPHIAN* (6,305 tons). Torpedoed and sunk by *U96* off North Arklow lightship (32 lost). Liverpool to Boston in ballast.

8.10.17 *GRELDON* (3,322 tons). Torpedoed and sunk by *U96* seven miles ENE of North Arklow lightship (28 lost). Out of Birkenhead for Italy with cargo of coal.

12.10.17 *W. M. BARKLEY*(569 tons). Torpedoed and sunk by *UC75* seven miles off East Kish lightship (4 lost). Liverpool bound with a cargo of stout from Dublin.

13.10.17 *ESKEMERE* (2,293 tons). Torpedoed and sunk by *UC75* fifteen miles WNW of South Stack, Anglesey (20 lost). Out of Belfast for Barry in ballast.

28.11.17 *APAPA* (7,832 tons). Torpedoed and sunk by *U96* NE of Point Lynas, Anglesey (77 lost). Liverpool bound from West Africa with general cargo.

30.11.17 *DERBENT* (3,178 tons). Torpedoed and sunk by *U96* off Point Lynas, Anglesey. Out of Liverpool for Cobh with cargo of fuel oil.

4.12.17 *HARE* (774 tons). Torpedoed and sunk by *U62* off East Kish lightship (12 lost). Out of Manchester for Dublin with general cargo.

7.12.17 *EARL OF ELGIN* (4,448 tons) Torpedoed and sunk by *UC75* in Caernarfon Bay (18 lost). Making for Dublin from London in ballast.

13.12.17 *STEPHEN FURNESS* (1,712 tons). Torpedoed and sunk by *UB64* west of the Isle of Man. Impressed as an armed boarding steamer.

15.12.17 *FORMBY* (1,282 tons). Torpedoed and sunk by *U62* in Caernarfon Bay (15 lost). Out of Liverpool for Waterford with general cargo.

17.12.17 *CONINBEG* (1,279 tons). Torpedoed and sunk by *U62* in Irish Sea (15 lost). Out of Liverpool for Waterford with general cargo.

24.12.17 *DAYBREAK* (3,238 tons). Torpedoed and sunk by *U87* off South Rock lightship (21 lost). Making for the Clyde with a cargo of pyrites from Huelva.

25.12.17 *AGBERI* (4,821 tons). Torpedoed and sunk by *U87* off Bardsey Is. Out of Dakar with general cargo for Liverpool.

27.12.17 *ADELA* (685 tons). Torpedoed and sunk by *U100* twelve miles NW of the Skerries, Anglesey (24 lost). Outward bound from Liverpool for Dublin with a general cargo.

28.12.17 *ROBERT EGGLETON* (2,274 tons). Torpedoed and sunk by *U91* ten miles SW of Bardsey Is. (1 lost). Making for Leghorn from the Clyde with a cargo of coal.

28.12.17 *CHIRRIPO* (4,050 tons) Struck mine laid by *UC75* off Belfast Lough. Outward bound from Liverpool to Kingston, Jamaica, with general cargo.

1918

5.1.18 *ROSE MARIA* (2,220 tons). Torpedoed and sunk by *U61* off North Arklow lightship (1 lost). Bound for Barry Roads from Scapa Flow in ballast.

6.1.18 *HALBERDIER* (1,049 tons). Torpedoed and sunk by *U61* off Bardsey Is. (5 lost). Manchester to London with general cargo.

26.1.18 *CORK* (1,232 tons). Torpedoed and sunk by *U103* off Lynas Point, Anglesey (12 lost). Making for Dublin with passengers and general cargo from Liverpool.

29.1.18 *ETHELINDA* (13,257 tons). Torpedoed and sunk by *U103* fifteen miles NW of Skerries, Anglesey (26 lost). For Barrow with cargo of iron ore from Bilbao (Spain).

29. 1. 18 *GLENFRUIN* (3,097 tons). Torpedoed and sunk by *UC31* NW of Holyhead (32 lost). Bound for Ardrossan with a cargo of iron ore from Serifos, Greece.

5.2.18 *TUSCANIA* (14,348 tons). Torpedoed and sunk by *UB77* off Antrim coast. Carrying US troops and general cargo from New York to Liverpool (44 lost).

5.2.18 *CRESSWELL* (2,829 tons). Torpedoed and sunk by *U46* off Kish lightship. Making for Gibraltar with a cargo of coal from the Clyde.

5.2.18 *MEXICO CITY* (5,078 tons). Torpedoed and sunk by *U101* fifteen miles WSW of South Stack, Anglesey (19 lost). Out of Liverpool for Alexandria with a general cargo.

11.2.18 *WESTPHALIA* (1,467 tons). Torpedoed and sunk by *U97* north of Dublin. Impressed as 'a special service ship'.

20.2.18 *DJERV* (1,527 tons). Torpedoed and sunk by *U86* off Skerries, Anglesey (2 lost). Out of Heysham for Newport, South Wales, in ballast.

23.2.18 *BRITISH VISCOUNT* (3,287 tons). Torpedoed and sunk by *U91* off Skerries, Anglesey (6 lost). Out of Liverpool for Queensland with cargo of oil.

26.2.18 *TIBERIA* (4,880 tons). Torpedoed and sunk by *U19* off Belfast Lough. Bound for New York with general cargo from the Clyde.

26.2.18 *DALEWOOD* (2,420 tons). Torpedoed and sunk by *U105* ten miles SW of Isle of Man (19 lost). Out from Cardiff to Scapa Flow with cargo of coal.

27.2.18 *LARGO* (11,764 tons). Torpedoed and sunk by *UB105* twelve miles W of Calf of Man. From Barry to Scapa Flow with cargo of coal.

1.3.18 *PENVEARN* (3,710 tons). Torpedoed and sunk by *U105* off South Stack, Anglesey (21 lost). Barrow to Barry Roads in ballast.

2.3.18 *CARMELITE* (2,583 tons). Torpedoed and sunk by *U105* ten miles SW Calf of Man (2 lost). Out of Bilbao for the Clyde with a cargo of iron ore.

2.3.18 *KENMARE* (1,330 tons). Torpedoed and sunk by *U104* off Skerries, Anglesey (29 lost). Sailed from Liverpool for Cork with general cargo.

7.3.18 *TARBETNESS* (3,018 tons). Torpedoed and sunk by *U110* in Caernarfon Bay 12 miles SW of Caernarfon lightship. Out of Manchester with general cargo.

17.3.18 *SEA GULL* (976 tons). Torpedoed and sunk by *U103* seven miles off Point Lynas, Anglesey (20 lost). Le Havre to Liverpool with general cargo.

20.3.18 *KASSANGA* (3,015 tons). Torpedoed and sunk by *U103* off South Arklow lightship. Out of the Clyde with a cargo of coal.

24.3.18 *ANTEROS* (4,241 tons). Torpedoed and sunk by *U103* sixteen miles W by N of South Stack, Anglesey (2 lost). Manchester to Port Talbot in ballast.

31.3.18 *CONARGO* (4,312 tons). Torpedoed and sunk by *U96* twelve miles off Calf of Man ((9 lost). Outward bound from Liverpool.

5.4.18 *CYRENE* (2,904 tons). Torpedoed and sunk by *UC31* in Caernarfon Bay off Bardsey Is. (24 lost). Out from the Tyne to Blaye (Gironde) with a cargo of coal.

14.4.18 *CHELFORD* (2,995 tons). Torpedoed and sunk by *UB73* ten miles NW by W of Bardsey Island. Clyde to Barry Roads in ballast.

16.4.18 *LADOGA* (1,917 tons). Torpedoed and sunk by *UB73* fifteen miles SE of South Arklow lightship (29 lost). Out from Bilbao for Maryport with a cargo of iron ore.

20.4.18 *LOWTHER RANGE* (3,926 tons). Torpedoed and sunk by *U91* off South Stack, Anglesey. Bound for the Clyde with a cargo of iron ore from Cartagena.

21.4.18 *NORMANDIET* (1,843 tons). Torpedoed and sunk by *U91* thirty-four miles SW of Calf of Man (19 lost). Out from Bilbao to Clyde.

28.4.18 *ORONSA* (8,075 tons). Torpedoed and sunk by *U91* twelve miles W of Bardsey Is. (3 lost). Bound for Liverpool from Talcahuano (Chile) and New York with general cargo.

23.5.18 *INNISFALLEN* (1,405 tons). Torpedoed and sunk by *UB64* east of Kish lightvessel (10 lost). Liverpool to Cork with general cargo.

5.6.18 *POLWELL* (2,013 tons). Torpedoed and sunk by *U96* off Dublin Skerries. Outward bound from Troon with a cargo of coal.

20.8.18 *BOLTONHALL* (3,595 tons). Torpedoed and sunk by *UB92* thirty-four miles SW of Bardsey Is. (5 lost). Out from Manchester for Gibraltar with cargo of coal.

21.8.18 *BOSCAWEN* (1,936 tons). Torpedoed and sunk by *UB92* off Bardsey Is. (1 lost). En route to Barry Roads from Birkenhead.

22.8.18 *PALMELLA* (1,352 tons). Torpedoed and sunk by *UB92* off South Stack, Anglesey (28 lost). Out of Liverpool for Lisbon with general cargo.

13.9.18 *M. J. CRAIG* (691 tons). Torpedoed and sunk by *UB64* off Belfast Lough (4 lost). Ayr to belfast with cargo of coal.

14.9.18 *NEOTSFIELD* (3,821 tons). Torpedoed and sunk by *UB64* off Co. Down coast. Out of Donaghadee and Clyde for Naples with a cargo of coal.

19.9.18 *BARRISTER* (4,952 tons). Torpedoed and sunk by *UB64* nine miles W by N of Chicken Rock lightship, Isle of Man (30 lost). Out of Glasgow for Liverpool and West Indies with general cargo and mail.

28.9.18 *BALDERSBY* (3,613 tons). Torpedoed and sunk by *UB91* off Codling Bank lightship (2 lost). For Avonmouth with a cargo of grain from Montreal.

10.10.18 *LEINSTER* (2,646 tons). Dublin–Holyhead mail/passenger ship torpedoed and sunk by *UB123* off Kish lightship (176 lost)

14.10.18 *DUNDALK* (794 tons). Torpedoed and sunk by *UB123* and *U90* off the Skerries, Anglesey (21 lost). Liverpool to Dundalk with general cargo.

18.10.18 *HUNSDON* (2,899 tons). Torpedoed and sunk by *UB92* off Strangford light buoy, Co. Down. Le Havre to Belfast in ballast (1 lost).

proceeded slowly along the Suffolk coast making for Harwich. Here they were moored in tidy groups of five. An uncertain number had been scuttled in German ports by their crews. On 21 November Admiral David Beatty accepted the surrender of the German Grand Fleet in the Firth of Forth. Llangefni's three year battle against the U-boat was officially at an end.

Back area repercussions

As they streamed home from the Racecourse after Gustav Hamel's demonstration flights at the end of June 1913, the Wrexham crowds, bitten by the aviation bug, could be excused for not anticipating the relatively little time left for flying for fun. Hardly had north-east Wales made its first acquaintance with the aeroplane, when, on the morning of 3 August 1914, the Home Secretary announced an immediate ban on all civilian flying. Aerobatic 'frivolities' were frowned upon, not in the least to save valuable machines from damage as their young, daredevil pilots turned to 'the grimmer business of engaging the Hun for the first time in aerial combat'. These were stirring words, typical of the sentiment of the time and ignored completely the country's unpreparedness for war, at least war in the air.

It was, perhaps, the night of 31 May 1915, when Zeppelin LZ38 bombed London's East End, sending shock waves throughout the Empire, that served as a catalyst and moved the Government to set its house in order. There had been seven previous Zeppelin raids, mainly on coastal towns, with minor damage and minimal loss of life. But now it was London's turn to be caught unawares, its air defences virtually non-existent. It had twelve AA guns, ten small, widely scattered fighter detachments, and a chaotic reporting system. In 1917 a new menace would emerge to supplant the Zeppelin, the huge Gotha and Staaken Giant biplane bombers.

Although Wales and the north-west were beyond the effective range of these bombers, the region was in theory, and occasionally in practice, within reach of the German naval airships almost until the end of the war. Zeppelin L21 crossed the Pennines on the night of 25/26 October 1916 and again on 27/28 November, but fortunately for Bolton, Macclesfield and Derby standards of night-time navigation did not match up to ambition or daring. On the night of 19/20 October 1917 Zeppelin L41 bombed Birmingham under the impression that it was over Manchester. On 12/13 April 1918 the captain of L61 for some reason aborted his attack on Liverpool whilst over Runcorn and, possibly attracted by the glow of blast furnaces, dropped his bombs on Ince-in-Makerfield, Wigan and Aspull before returning to Germany via Radcliffe, Oldham and Royton, successfully evading the Great Yarmouth-based Fairey F2A seaplanes sent up to intercept it.

Zeppelins also penetrated the Black Country where the Lucas, Wolseley and Austin aircraft and component factories offered legitimate strategic targets. On the night of 31 January/1 February 1916 Zeppelins L19 and L21 raided Birmingham, the latter, under the command of Kapitanleutnant Max Dietrich, by default, since he thought he was bombing Birkenhead docks and Liverpool! There were two other sorties over the Black Country and eastern Shropshire, on 19/20 October 1917 and 12/13 April 1918.

But all these were relatively late incursions. The Government's initial response to these new threats and the subsequent upsurge in aerial activity focussed on the Home Counties and the Channel coast and in the early months of the war had little impact on North Wales and the north-west. As was traditional, the war effort in these areas centred upon weapons production and recruitment and training for the Royal Navy and Army, especially the local battalions of the RWF, the KSLI, the Cheshire Regiment, and the county Yeomanries. In aviation terms this isolation was not to last. On 4 February 1915 Germany officially declared the Irish Sea and coastal waters a war zone, with merchant shipping liable to be sunk on sight. But two weeks before this

announcement a German submarine had penetrated into Liverpool Bay, sinking three ships off the port, with a further two in early February, all possibly by the same submarine.

In the Zeppelin-free zone on the eastern side of the Welsh mountains things were also starting to move. Even before the outbreak of war aircraft manufacture had progressed in the short space of five years from the two-man workshop, feeling their way by trial and error, to a mass-producing industry of large companies. The opening of hostilities gave the nascent industry an even greater boost, only to be followed by abrupt cut-backs and cancellation of orders in the pipeline after the Armistice in November 1918.

Avro's Manchester factory was already churning out 504s. No. 2 National Aircraft Factory, forerunner of Fairey Aviation, was built at Heaton Chapel while Alexandra Park was requisitioned as 15 AAP (Air Acceptance Park) where aircraft were test flown and stored. A sub-storage depot was also formed at Hesketh Park, Southport, using the sandy foreshore as a convenient landing-strip.

On Merseyside No. 3 National Aircraft Factory was built alongside Aintree Racecourse — a ready-made flight testing ground — primarily to produce under licence the Bristol Fighter, affectionately known as the 'Biff' or 'Brisfit', one of the mainstays of the RAF in its formative years.

In the Midlands Castle Bromwich flying ground had been requisitioned in 1914 and a Training Squadron established. Numerous reserve squadrons would also be formed here. In July 1918 No. 14 Aircraft Acceptance Park was embodied on the airfield, testing locally built Handley Page 0/400 twin-engined heavy bombers and S.E.5s, the latter to prove one of the outstanding single-seat fighters of the war. In practice never more than a sub-depot of Castle Bromwich was AAP Monkmoor (unnumbered) on the tiny flying field at Monkmoor within a loop of the Severn in the eastern suburbs of Shrewsbury. The latter's First World War hangars and some of its ancillary buildings managed to survive the encroaching housing and hospital development of the 1920s, which explains why, during the Second World War, Monkmoor would re-emerge as the base for No. 34 Maintenance Unit, responsible for recovering aircraft wrecks throughout North and Mid-Wales and the north Midlands.

The more familiar pattern of military aviation along the Welsh borderland during First World War I was first sketched in some twenty-two years earlier with the laying out of airfields in the so-called 'back areas' for Training Squadrons (TS) or Training Depôt Stations (TDS) at Shotwick (Sealand), Hooton Park, Tern Hill and Shawbury. Only the first would survive post-war defence cuts, but all were revived between 1936-39, and, complemented by further operational and training aerodromes, would intensify even further the levels of aerial activity in Welsh border skies.

The inhabitants of the North Wales coastal strip had very quickly become accustomed to seeing Llangefni's blimps making a welcome landfall and hugging the coastline home. Apart from the actual physical dislocation caused by new aerodromes and the traumatic social impact they had upon adjacent rural communities, the advent of military aircraft in ever increasing numbers impinged itself upon the sensibilities of people along the northern March of Wales in three main ways.

The first was obvious, blunt and brutal. Aircraft — and occasionally bodies — began to fall out of the sky in increasing numbers. Forced landings were made willy-nilly, breaking down for ever the barriers of rural isolation and, if not actually according some sort of prestige to a community, providing a new dimension to pub gossip and establishing another bench mark to the oral archive of parish pump history. Thus on Sunday, 10 March 1918 the air war touched the south Cheshire village of Malpas. Its inhabitants had waited a long time for their fleeting moment of glory.

For six months they had had to make do with a second-hand

The regimented standard CWGC headstones stand out in many a border churchyard. These WW1 graves are at Shotwick, looking south over the former aerodrome at Sealand.

'aeroplane mishap' that had occurred on Tuesday, 2 October 1917, when a 2/Lt. A. C. Tallent, 10 TS Shawbury, well off course and with failing engine, had made an emergency landing at Four Lane Ends, Overton-by-Malpas, crashing through a hedge, 'breaking a propeller and damaging the sails'. The pilot scrambled out, visibly shaken but otherwise unhurt. A boy was dispatched on horse-back to Malpas police station. But as he sat in a farmhouse kitchen, fortified by a large glass of whisky, neither Tallent nor the curious crowd of well wishers, suspected that four weeks later he would be dead, killed outright in another flying accident. On Sunday, 4 November 1917 the engine of his Sopwith Camel (serial B6252) cut whilst making a turn and spiralled out of control into the ground.

But now the *Whitchurch Herald* sought to gain maximum news value from 'the first flyer to land in Malpas'. In the untutored language of the lay hack 'the pilot had intended to bank, but something got out of gear and he came perilously close to collapse'. In other words the engine had stalled. The stricken plane swept under the telephone wires in Old Hall Street, sliced off the tops of some standard roses in a cottage garden and landed in a field out towards Bradley. The pilot must have been one of a group exercising together, for almost immediately he was followed down by 'another Canadian pilot and observer' who landed at Old Hall and went to his assistance. Fortunately he was uninjured and, swapping places with the observer, was flown back to base, leaving the crippled plane to be collected by a salvage team from Shawbury.

This would be a scenario repeated many times. Pages 74–78 list some 120 accidents in the area during the last two years of the war. Considering the precarious nature of surviving official records even this tentative catalogue must be considered as only the tip of the iceberg!

A natural corollary of these, and another means whereby people became aware of the darker side of the air war, was the brash intrusion of the military funeral into the normally placid regimen of the country church. As casualties mounted a special plot was set aside in the churchyard or cemetery, immediately recognisable by the standard white IWGC headstones standing to attention, perfectly spaced and carrying regimental insignia and briefest of biographical details. Ranks are broken only where perhaps grieving relatives were quick to erect their own memorial stone or a body has since been exhumed to be returned to the dominion or colony from whence it came.

One could pause at the RFC/RAF graves in Shotwick, Shawbury,

Eastham and Stoke upon Tern churchyards, designated burial grounds for casualties from nearby training aerodromes. But because of their compactness interest focuses momentarily on Tilstock's tiny churchyard and cemetery. Strangely, bearing in mind the proximity of RAF Tilstock (on Prees Heath) and its satellite at Sleap, there are no Second World War burials. In the last conflict training and operational casualties were so high that a special CWGC cemetery was opened at Blacon near Chester which eased the pressure on parochial burial grounds in the region. But in the First World War Prees Heath was home to a massive Army camp and its hospital gave succour to injured airmen from both Tern Hill and Shawbury if they crashed in the vicinity. With equal facility its mortuary held the burnt and broken bodies of unfortunate aircrew until the inquests were held and arrangements made for burial, usually, at the behest of relatives, in someone's home town. If this were not possible — no known kin or the dead person was a colonial — Tilstock cemetery became the receiving burial ground. There are nineteen military gravestones (with possibly five removed) in Tilstock's closed churchyard covering the period 1 March 1916–18 June 1917 with significant concentrations of burials suggesting epidemics of some sort at Prees Camp. With effect from July 1917 all military burials took place in the new cemetery across the road, where there are another nineteen First World War graves (with at least one removed). Of the ten graves of aircrew at Tilstock, four are Australian Flying Corps, two Canadian and one South African. The list reads:

> 2/Lt. E. P. Hughes, RFC. 27 July 1917.
> 959 Cadet E. J. C. Treadwell, AFC. 20 September 1917. Aged 22.
> 1397 Gunner W. H. Herford, AFC. 4 October 1917. Aged 49.
> 2/Lt. J. F. Kneale, RFC. 21 December 1917.
> 1130 Corp. A. Morgan, AFC. 10 February 1918. Aged 31.
> 2/Lt. R. J. T. Forsyth, AFC. 16 February 1918.
> 29412 1st Air Mech. C. R. Clack, RAF. 5 May 1918.
> Lt. W. B. Bickell, RAF. 12 October 1918. Aged 25.
> Flight Cadet George L. Robinson, Princess Patricia's C. L. I. Att'd RAF. 1 November 1918. Aged 21.
> 2/Lt. G. P. Cilliers, RAF. 10 November 1918.

Expanding on the basic information inscribed in stone, the researcher discovers no great calamity, just a typical selection of the everyday hazards and dangers that beset would be aviators learning to fly in none too reliable machines. 2/Lt. Edward Phillip Hughes of 131 Squadron, Shawbury, was a native of Ellesmere, Cape Province, 'a gallant young soldier' who had distinguished himself in the fighting in German East Africa, gaining a commission and a transfer to the RFC. His tired Airco DH2, a 'pusher' biplane serial 6008, although relegated to a training role, was still a tricky aeroplane to fly on account of its very sensitive controls and difficulty in recovering from a spin. It crashed on 9 June 1917. Hughes survived, albeit with terrible injuries to legs and lower body as the 100 h.p. Gnome rotary engine joined him in the cockpit. He died in Prees Military Hospital on 27 July, shortly after his brother had arrived from South Africa. He was buried with full military honours, the first such funeral in Tilstock's new cemetery and the twentieth in the village church. However, his headstone has since been removed following exhumation after the war and the transfer of his remains to his native country.

F/Cdt. Edward Jabez Cooper Treadwell, 30 TS (Aust.), Tern Hill, hailed from St. Kilda, Victoria, and was in the first weeks of training. A group of cadets had been standing on the wing of an aircraft just about to take off, watching pre-flight checks and starting procedures. As he jumped off the wing he walked straight into the propeller, a tragic, messy end to a short patriotic life. Four days before Christmas 1917, 2/Lt. John Francis Kneale, 10 TS, Shawbury, flying a Sopwith Camel B2310, became lost in thick fog somewhere south of Whitchurch and crashed when he descended to try and check his position.

2/Lt. Reginald James Thomas Forsyth, 6 TS AFC, Tern Hill, was another Australian who succumbed to flying injuries in Prees Military Hospital. Catching up with days lost due to bad weather he had gone up with others on Sunday, 20 January 1918, to practice formation flying. He found himself slightly ahead of the rest and in order to regain position made a vertical bank during which he lost control and went into a spiralling nose dive from 1,000ft and crashed. He sustained injuries to head and legs that would prove fatal. He died on Saturday, 16 February 1918.

Air Mechanic (1st Class) Charles Richard Clack, 131 Squadron, Shawbury, was killed on 5 May 1918 when acting as observer to Lt. D. M. Miller in an Airco DH6 C6657. They experienced engine failure and in a heavy forced landing Clack was thrown out and killed. His pilot was only slightly injured. The aircraft obviously faired slightly better and was repaired, ending its operational career with 254 Squadron at Prawle Point, Devon, on coastal reconnaissance and anti-submarine work.

Lt. William Burt Bickell, 13 TDS, Tern Hill, hailed from Toronto and was on attachment from Princess Patricia's Canadian Light Infantry. He has two memorial stones at Tilstock — the official IWGC headstone, and a 'civilian' stone erected by his mother which also commemorates his brother, C. T. Bickell, also serving with PPCLI, and killed in action in Belgium. From Montreal and seconded from the PPCLI and Eastern Ontario Regiment, was F/Cdt. George Lancaster Robinson, 13 TDS. He died at Prees Military Hospital on 1 November 1918 of pneumonia, complications following an earlier flying accident.

The threat of air raids was something new for civilians and an eventuality much to be feared before the fighting was over — only the military were supposed to die in a war fought in France and Belgium. This and the accidents, forced landings and regimented headstones in the local churchyard made people uneasily aware of an air war and coloured their perception of the RFC/RAF. A third dimension was added as provincial newspapers began to carry stories of local RFC/RNAS men promoted, decorated, killed, wounded, missing in action, POW on the Western Front, in Italy and the Middle East — not as many as in the Army or Royal Navy for the ratio of, for example, Wrexham men serving in the three forces was 65:1¹/2:1. This is seen repeatedly when in the 1920s casualty lists were synthesised and neatly tabulated on countless parish war memorials. Unfortunately for the local historian there are too many memorials where names are not differentiated according to service, regiment or rank. The small village of Gresford, between Wrexham and Chester, has fifty names on its First World War Roll of Honour in the porch of All Saints' Church, all Army (twenty-four RWF) except for two RN/RNVR and two RFC/RAF. The small market town of Ruthin, at the upper end of the Vale of Clwyd, carries seventy names (including thirty-six RWF) but only two RAF/RNAS. This particular memorial highlights the haphazard and often unco-ordinated manner in which such monuments were erected. On-going research has since revealed a further twenty nine names (twenty-three RWF) that ought to have been included on the stone!

The fallen from the borough of Wrexham are commemorated in the War Memorial Chapel at the east end of the north aisle the parish church of St. Giles. It was fitted up in 1918–19 by Sir Thomas Graham Jackson who had been responsible for the part refurbishment of the chancel in 1914. The nominal roll on five bronze tablets set in stone tablets contains some 337 names, 325 Army (173 RWF), seven Navy and but four RFC/RAF — D. W. Cartwright, P. G. Harris, A. H. L. Soames and S. B. Welch. Strict censorship in force at the time makes it difficult to discover how exactly these men died and the task is complicated by the fact that details released to the press by relatives are inevitably at variance with official records.

Because he was killed relatively early in the war and possibly reflecting his higher social standing, Captain Arthur Henry Leslie Soames, 3rd King's Own Hussars and RFC, scion of the Wrexham brewing family, also has an individual bronze plaque on the tower arch in the 'Soldier's Chapel' at the west end of the church. On 13 August 1914 he had been one of the first airmen to risk his machine on the Channel crossing and within three days of his arrival at Maubeuge, GHQ of the British Expeditionary Force, was carrying out the first aerial reconnaissance of the war. For his daring and work in convincing Army staff that aircraft and balloons were equally effective, if not superior, instruments of intelligence gathering as horse and foot patrols, Captain Soames was mentioned in dispatches, gained the French Légion d'honneur and was awarded the Military Cross. June 1915 saw him as instructor and squadron commander at the Central Flying School at Upavon, Wiltshire, established in June 1912 under the command of Captain Godfrey M. Paine, RN, and now busy churning out a steady stream of pilot cannon fodder for the voracious front-line squadrons in France.

Captain Soames also flew for the Experimental Flight, evaluating new equipment. On Wednesday, 7 July 1915 he was involved near Netheravon in the testing a new type of HE bomb detonated by fuse. According to one report the bomb exploded prematurely on his aircraft during or shortly after take-off. He suffered terrible injuries; his shattered left arm was quickly amputated but to no avail. He died the next day. His family received a letter of sympathy from King George V who had been impressed by Capt. Soames's exhibition of 'bomb throwing' at Upavon during a recent visit and clearly recalled that officer's investiture with the M.C. at Buckingham Palace. On the

RFC wings head the inscription on the grave of Capt. A. H. L. Soames of the Central Flying School, tragically killed in July 1915.

motion of Councillor S. G. Jarman, mayor, a full meeting of Wrexham Borough Council on the Friday evening passed a 'vote of condolence' expressing the sympathy of 'all classes' of townsfolk. Strangely, although Capt. Soames lies buried in Gresford churchyard his name does not feature in the War Memorial north porch of All Saints, erected in 1920–21, a design by Sir Thomas Graham Jackson (as was that at Wrexham). In the same year the latter was also responsible for the refurbishing of the north-east chapel in St. Mary's, Mold, again as a War Memorial.

Air Mechanic (3rd Class) David William Cartwright's home was in Saxon Street Wrexham. He was attached to 55 Kite Balloon Section. The RFC operated captive balloons as observation platforms. They differed from free balloons in that they were sausage-shaped, modelled on the German *Drachen*, and had rudimentary rudder, wind sail, and drogue streamer flying from the tail to stabilise them. Along with twenty-three others Cartwright was drowned when, on 31 December 1917, the Fleet Messenger *Osmanieh* (4,041 tons) was either torpedoed by the German submarine *UC34* off Alexandria harbour or struck a mine just laid by that U-boat — the reports are conflicting. Their names are recorded on the Chatby Memorial, Egypt.

Old Wrexhamian 2/Lt. Percy George Harris had been commissioned into the RWF (17 Batt.) but had transferred as an observer to the RFC. On Saturday, 11 August 1917 he had taken off at 18.25 hrs. from La Lovie on a photo-recce mission over Ypres in an RE.8 (serial A3863) of 21 Squadron, piloted by 2/Lt. C. E. Holoway. At about 19.10 hrs. they met up with Leutnant T. Quandt of *Jagdstaffel 36*, a single seater fighter unit, and were shot down west of Zonnebeke. They have no known grave but are remembered on the Arras Memorial. Harris's death was widely lamented in Wrexham, where he had been a teacher at the National Schools, and organist at St. John's Church, Hightown.

Lt. Stanley Birks Welch, 49 Squadron RAF, whose parents lived in York Street, Wrexham, was killed in action on Sunday, 25 August 1918, along with his observer 2/Lt. David Charles Roy. They had taken off at 16.05 hrs. from Bourton Wood on a bombing mission, flying Airco DH9 C6209. It was a retaliation daylight raid for the night bombing of Bertangles aerodrome by the Germans but everything went wrong from the start. 49 Squadron was making for Etreux, HQ of *1 Bombengeschwader*. They were hampered by a sudden thunderstorm and their promised escort, the SE5as of 32 Squadron, tangled with a formation of Fokker D.VIIIs and did not make the rendezvous. C6209 was engaged by an enemy aircraft and had its tail planes shot away. It was last seen going down in a controlled spin over Moivres before diving into the ground. Both aircrew are buried at Vis-en-Artois cemetery.

Of the airmen on the Gresford memorial, Capt. A. R. James, 62 Squadron, ex-24th Battalion (Denbighshire Yeomanry) RWF, was killed on 24 March 1918, a day when the RFC lost some forty-seven aircraft as low-flying patrols and bombing missions were ordered from every available machine that could be put into the air. On the ground German forces had advanced fifteen miles in three days and the Army was desperately trying to stall the enemy advance whilst new defences were feverishly erected. Despite his squadron being in the throes of a move northwards from Cachy to Remaisnil James and his observer, Lt. J. M. Hay took off at 16.20 hrs. in their Bristol F2b fighter for a spot of trench-strafing at Peronne-Ham. They were shot down by enemy machine-gun fire. Lt. Hay was killed outright. Capt. (Lt. in casualty report) James must also have been seriously wounded. He managed to land his plane behind enemy lines and it was first thought that he had been made a prisoner of war. Whatever, he obviously succumbed to his wounds.

The second Gresford airman was 20-year old Capt. William Victor Trevor Rooper, formerly of the Denbighshire Yeomanry and 24th Battalion RWF, and since mid-September 'B' Flight commander, No. 1

Squadron RFC, based at Bailleul (Asylum Ground) since March 1917. On 27 September the squadron had taken delivery of its first Nieuport 27 fighter, an over-rated, under-powered aircraft, one however which would find a ready market in the American squadrons in the latter stages of the war. But 1 Squadron would not take delivery of the SE5a until January 1918. So it was in a Nieuport 27 serial B6767 that Rooper took off at 14.50 hrs. on 9 October 1917 on a low-level offensive patrol in support of the infantry surge in the salient between Yser and Houthulst Wood. It was raining, deadly enemy of troops on the ground but by now no reason for aircraft to remain on the ground. Air support was almost expected as of right on the day of an assault and No. 1 Squadron laboured on gamely. Over Polygon Wood Capt. Rooper met up with five enemy aircraft and was shot down by Leutnant Dannhuber of *Jagdstaffel 26* and was forced to land, crashing into the British front-line trenches, sustaining, amongst other wounds, a fractured thigh. He died of these injuries on 25 October 1917. At the time of his death Capt. Rooper was something of an 'ace' having claimed eight enemy aircraft, four of them in B6767. Capt. Rooper's 'score sheet' reads as follows

1917	Nieuport	Location	Enemy Aircraft
28 July	B1675	Becelaere	Albatros D.V — out of control
9 Aug.	B1675	Houthoulst	DFW.C — captured
17 Aug.	B1675	Tenbrielen	DFW.C — out of control
11 Sep.	B33632	Houthoulst	Albatros D.V — out of control
19 Sep.	B6767	Poelcapelle	Albatros D.V — destroyed in flames
25 Sep.	B6767	Gheluvelt	Albatros D.V — destroyed
1 Oct.	B6767	Houthoulst	DFW.C — out of control
5 Oct.	B6767	Zandvoorde	Albatros D.V — destroyed

The two airmen on the war memorial in Wynnstay Road, Ruthin, came from different social backgrounds, the one on direct entry commission into the RAF and relatively inexperienced, the other an old hand who had worked his way up through the ranks as it were. F/Lt. John Emyr Thomas, RNAS, aged 23, a product of Ruthin Grammar School and Grove Park School, Wrexham, had been a premium apprentice with Singers, Coventry. He enlisted in the Royal Naval Division in September 1914 and had seen active service at Antwerp and as a dispatch-rider at Gallipoli, where he had been wounded. In April 1916, after hospitalisation in Malta, he was commissioned into the RNAS, carrying out fighter and reconnaissance missions over the Somme with 4 Wing before being posted to No. 2 AAP Hendon, test flying aircraft received from the manufacturers before ferrying them out to aircraft parks/depots with the Expeditionary Force. He also flew in the *ad hoc* AZP maintained at Hendon. F/Lt. Thomas was killed on 3 September 1917 in a flying accident at the new airfield at Yeovil where he had gone to pick up an Avro DH4 from the Westland Aircraft Company. He was buried with full military honours at Rhewl chapel on 7 September, a guard of honour being supplied by NCOs and men from the Yeovil Volunteers and fellow officers of the RNAS.

For 2/Lt. D. B. Griffith, 209 Squadron, then based at Bertangles, near Amiens, it was third time unlucky. He had already survived two forced landings, one on 6 July 1918 in Sopwith Camel D3328 which landed at Pont Noyelles and had to be dismantled *in situ* since the ground was too boggy for take-off, and a second the following day in Camel C124 which overturned on landing. 2/Lt. Griffith, aged 19, was posted missing on 19 July 1918, lost on a 'special mission'. An Allied counter-attack launched on 18 July had brought the German advance west of Rheims to a standstill which developed into a rout. Enemy lines of communications, aerodromes and ammunition dumps became the targets. 2/Lt. Griffith was last seen over German-held Cappy aerodrome at 04.45 hrs. when his Camel (C193) was hit by machine-gun fire from *Flakbatterie 82*. He was initially posted as 'missing in action'. Strangely the full story of his death did not emerge until January 1919, possibly because the next day, 20 July 1918, the squadron had shifted base to Quelmes and during the confusion fell behind with its paper work.

A letter from the squadron CO, Major J. O. Andrews D.S.O., M.C., was more enlightening than the usual terse Air Ministry telegrams: 'At the time we were bombing and shooting up an enemy aerodrome at dawn, this being carried out at 200 ft. Fifteen machines and pilots started out and two did not return, one being 2/Lt. Griffith, the other 2/Lt. Scadding, the latter having been reported a POW. This work we were engaged on was of a very hazardous and perilous nature and I am of the opinion that this officer was shot down and killed by machine-gun fire from the ground. The aerodrome we were attacking was at Clappy on the Amiens-Albert front and we were stationed at Bertangles'. Lt. E. Scadding's mount was Camel D9629, brand new and on squadron strength for only a fortnight. Its destruction is credited to *Flakbatterie 764*. By coincidence Scadding had force-landed Camel D3328 on 3 July 1918, four days before Griffith's mishap in the same aircraft. The International Prisoner of War Society at Geneva later confirmed that 2/Lt. Griffith was brought down east of Hamel and was buried there. Today Lt. Griffith lies buried in the Villers Bretonneux Military Cemetery, south-east of Amiens, plot 6, row C, grave 10. In St. Peter's Church, Ruthin, there is a further memorial to Griffith — a rather fine painting by his artist sister, Mignon Griffith, depicting the three Marys at Christ's sepulchre set against a backdrop of the Clwydian Range.

But it is perhaps the name of Old Wrexhamian 2/Lt. G. B. Bate, RFC, the only airman on Caergwrle's war memorial, that encapsulates the steel nerve, the raw, stark courage and short life expectancy — two weeks on average — of aircrew over the Western Front. In the ferocity of the fighting the carnage in the air matched that on the ground. Yet as they flew steadfastly to almost certain death, obeying the rather silly orders to 'stand and fight', they were borne up by a strong faith and the sure knowledge that they were simply 'doing their duty'. On 23 April 1917 2/Lt. Bate was wounded, six days later he was dead.

On 23 April the RFC/RNAS lost some 42 aircraft, a high total even for those dark days. There were many enemy machines operating over the battlefield and there was intense skirmishing and dog-fighting. In the early evening, at about 1730hrs, 2/Lt. Bate was observer/gunner in a FE2b (serial A823) of 18 Squadron, then based at Bertangles. It was on a bombing mission and strictly speaking should have had an escort of Sopwith Pups, but these were too busily occupied with two formations of Albatros and Halberstadt scouts. A823 was attacked over Barelle by four enemy aircraft one of which was hit several times by Bate and sent out of control into a spinning nose dive. However, almost immediately his own radiator was shot to pieces, his pilot, 2/Lt. E. L. Zink badly wounded and he himself taking a flesh wound in the shoulder. Somehow A823 managed to make it back to base. At 14.20 hrs. on 29 April Bate was back in the air on escort duty in FE2b A5483. Unfortunately they met up with *Jagdstaffel 11*, which had just carved up 12 Squadron and proceeded to do the same with 18 Squadron. A5483 was shot down by *Leutnant* K. Wolff, *Jasta 11*, over Pronville at 16.00hrs. 2/Lt. Bate was killed but his pilot 2/Lt. G. H. Dinsmore fortunately survived unscratched, to fight another day.

Border aerodromes

While headstones, brasses, stained glass and war memorials yet survive to point to distant conflict in the air, the First World War aerodromes that spawned such aerial activity experienced mixed fortunes. As already noted, 244 Squadron's base at Glan-y-mor-isaf disappeared without trace. It took just five months to clear the site completely, the Bessoneau hangars being the last to go in June 1919.

Without exception these early aerodromes were all-grass airfields with hutted or tented accommodation, and their obliteration and reversion to farmland was achieved without much trouble or disfigurement of the rural landscape.

There appears to be no particular reason behind the selection of sites for training aerodromes in north-east Wales and adjacent Cheshire/Shropshire border country unless it was the obvious — relatively flat, obstruction free and capable of being upgraded with minimal engineering works and at minimum cost. Pre 1914 Hooton Park was a racecourse, laid out sometime after 1875 when the Naylor family, a Liverpool banking dynasty, left Hooton for their other seat at Kilmarsh, Northants. The aesthetic integrity of their Wirral estate had been compromised by the building of the Manchester Ship Canal. Humphrey Repton's landscaped parkland proved an ideal setting for a racecourse, complete with grandstands below the east front of the house and a polo ground in the centre. On 4 August 1914, the day after the last race meeting, Hooton Park, like so many other wide open spaces, was requisitioned by the Army for training and billeting purposes. A rifle range and assault course were laid out and the grounds littered with practice trenches and bunkers. The empty hall was taken over as a military hospital.

But following the entry of the USA into the war on 6 April 1917, plans were made for the assembly and storage of American aircraft imported via Birkenhead and Liverpool and for the training of Canadian and US pilots. Aerodromes and Air Acceptance Parks at Sealand and Hooton Park were earmarked for the purpose and construction work started. In July 1917 the army moved out of Hooton Park and Cubitt Ltd. moved in to erect three double Belfast General Service aeroplane sheds (still in situ, two with Listed Building status) and a single ARS shed (demolished in the 1930s) to the west of the racecourse. Hooton Hall now became the Officer's Mess, and structurally never recovered. The outbuildings at the Home Farm were commandeered for ORs' accommodation. Two further aeroplane sheds were planned to the north, but this would have meant the demolition of 'The Priory', gardens, and much of the Long Plantation. This particular scheme was not proceeded with, at least not until 1940 when a B1 hangar was built. The latter still stands and is used as a warehouse by a haulage contractor. But even before the builders had finished plans for an aircraft storage park were scrapped and the facility re-designated a Training Depôt Station, albeit retaining a 'half Aeroplane Repair Section'. The comments of 84-year old Col George Mousley, late Denbighshire Hussars and former agent to R. C. Naylor, on the felling of woodland at Hooton Park to make room for the airfield are best left unprinted! The flying ground was contained roughly within the racecourse fences, but was sufficient to give maximum runs of 400–500 yards, making allowances for clearance of surrounding patches of woodland.

On 19 September 1917, although the aerodrome was far from complete, No. 4 Training Depôt Station arrived from Tern Hill with its Sopwith Scouts and Dolphins. It was intended that 4 TDS should comprise three training squadrons but plans were slow to crystallise as building work would not be completed until mid-August 1918, with final hand-over date being 1 September 1918. Upon Hooton Park's closure in 1919, 4 TDS would move to Shotwick (Sealand) as a Training Squadron or Training School. Here, in April 1920, it disbanded and immediately reformed as 5 FTS (Flying Training School), whose aircraft were to be numbered amongst some of the more spectacular crashes in the north-east borderland, and which, by using Borras Lodge farm as a Relief Landing Ground (RLG), confirmed the potential of those fields as the future site of RAF Wrexham. In St. Mary's Church, Eastham, there is a wooden board inscribed 'In memory of airmen who lost their lives whilst training for their country' — most from 4 TDS, Hooton Park, namely, five Canadians, two Englishmen, and one American. The latter was

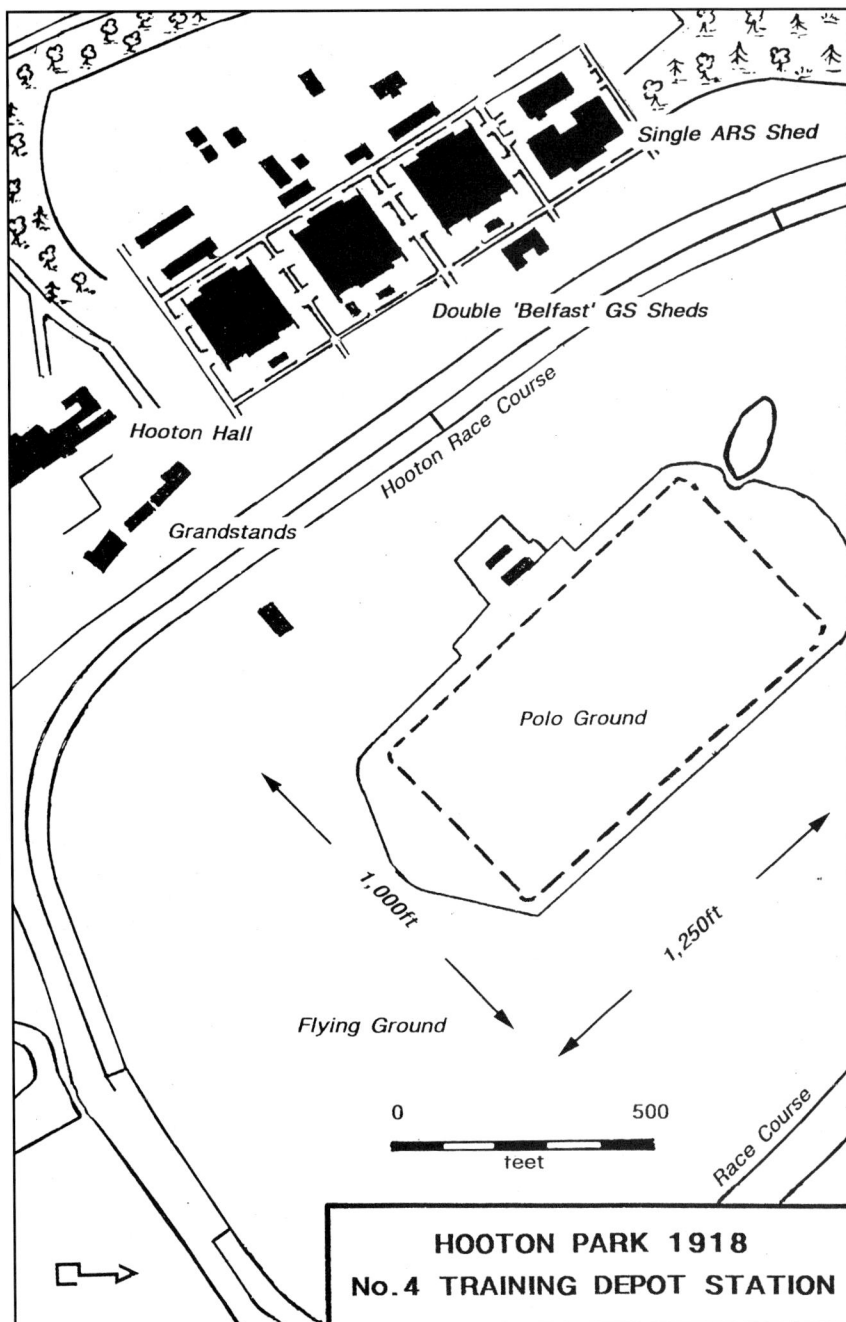

HOOTON PARK 1918
No. 4 TRAINING DEPOT STATION

salt marshes, housed both the Vickers testing ground, and No. 10 (followed by No. 63) Reserve Squadron, with which between 22 March–12 July 1917, an upwardly mobile Lt. McCudden had temporarily filled the new post as Wing fighting instructor. Although the FB.16D and its reincarnations as FB.16E and FB.16H produced the highest speeds by a fully armed fighter, this was only possible by using foreign engines such as the Hispano-Suiza and it did not go into general production. There was great sadness at Hooton when they heard of the death in a flying accident of Capt. (just promoted Major) McCudden whilst ferrying an SE5a (serial C1126) out to 60 Squadron at Boffles.

Hooton Park's technical site was only 70ft above sea level and the airfield sloped gradually down to the Manchester Ship Canal and the estuarine mud flats of the Mersey less than a quarter of a mile away. With the help of boatmen from Eastham Ferry, Whitby Locks and Garston Docks the aerodrome maintained a rudimentary air sea (more correctly river) rescue service since their aircraft — and those of Shotwick regularly made forced landings in the estuary between Bromborough and Ince Marshes where large areas of sandbanks and mudflats were exposed at low tide. This empty space was a favoured area for aerobatics and low-level flying.

The 'emergency services' are seen working together on Sunday, 27 January 1918. A Shotwick trainee, 2/Lt. John Daniel George Brendel, aged 22, Canadian, was flying a 'scout machine' (possibly a Sopwith Dolphin) at 6,000ft over the estuary at Eastham. On the evidence of a lock-keeper on the Manchester Ship Canal 'reports were heard from the engine like the reports of a revolver' and the plane went into a spinning nose dive. The duty officer at Hooton Park had already noticed something wrong and thought that the pilot was practising nose dives or had fainted in his seat. Whatever, the aircraft crashed into the estuary and sank, with just the tip of the propeller showing above the water. Lt. F. J. Thompson from the aerodrome went out to the sandbank by boat as the tide went out and recovered the body, still strapped into its seat. A Mersey Docks & Harbours Board lighter later recovered the wreckage.

In a later war, at almost the same spot, the Mersey would claim other aircraft. On 10 October 1940 it was Hawker Hurricane L1547, the first production model now operational with 312 (Czech) Squadron at Speke after nearly three years in the hands of the makers and the AAEE, Martlesham Heath. 312 Squadron had been working up at Speke since 26 September and was declared operational on 2 October. Its task was to provide daylight defence of Merseyside. On Thursday, 10 October the engine of L1547 caught fire during a map-reading familiarisation flight over the area and plunged into the Mersey. Sgt. Otto Hanzlicek baled out too late and too low. Before the New Brighton lifeboat could be launched - a cumbersome, long-winded process involving Flag Officer, Mersey for air-sea rescue purposes - Hanzlicek was swept away by the tide and drowned. His body was recovered on 1 November at Widnes.

In August 1997 the news broke that divers using latest techniques and equipment had possibly located L1547 at the bottom of the Mersey, by first reports well preserved by the river's silts and muds. The Ministry of Defence were awaiting sanction from Liverpool City Council to begin a salvage operation. However, it must be remembered

probably serving with a detachment of 43 Aero Squadron USAS with its Airco DH4s, farmed out from South Carlton (Lincolnshire) for two months before being sent to France in October 1918. An even shorter serving lodger unit was 117 Squadron, a day bomber unit raised at Waddington on 1 January 1918 but which did not become operational before the end of the war. In March 1919 it was sent to Ireland for internal security duties equipped with DH9s. It stopped at Hooton Park for ten days, 8–18 March, gathering its strength and servicing its machines before venturing out across the Irish Sea to Tallaght, near Dublin.

An interesting visitor to Hooton Park in May 1918 was Capt. J. T. B. McCudden, V.C., D.S.O.*, M.C.*, M.M., not only because he was the most famous British 'ace' with fifty-seven enemy aircraft to his credit, but also because he was flying the distinctively coloured (red overall) Vickers FB.16D (serial A8963), a one-off ever evolving fighter prototype which he liked to fly on his visits to Joyce Green (Kent) whilst on leave from the Front. This aerodrome, on the Dartford

that there are several other Second World War wrecks in the Mersey mud, several of them from botched launchings from the MSFU catapult at Speke.

In many ways 2/Lt. Brendel and Sgt. Hanzlicek had much in common — both rallying to wars against the German aggressor, one a patriot from the Dominions, the other in enforced exile from his homeland. After the occupation of Czechoslovakia Hanzlicek had escaped via Poland to France where, after training, he flew with Groupe de Chasse II/5 (a pursuit or fighter squadron) at Toul-Crox-de-Metz. In the shambolic period 10 May 25 June 1940 Hanzlicek claimed the destruction of a Dornier Do17, a Heinkel He111 and, a Messerschmitt Bf109 before escaping to England via North Africa and Gibraltar, learning on the way the Croix de Guerre and the Czech War Cross.

After it was relinquished by the RAF in 1919 Hooton Park reverted to agriculture with the smaller buildings used for light industry. 'Given over to solitude, its landing ground overgrown, its hangars empty, it seemed doomed to decay, its facilities and advantages forgotten or ignored' thus it was described by the *Mersey Air News* in 1929. But change was imminent. Hooton Park would re-emerge in the aviation world in the late 1920s with small-scale aircraft assembly, a short-lived stint as Liverpool's municipal airport, and as the base for Liverpool and District Aero Club, whose Avro Avians paid regular weekend visits to the Borras Lodge (Wrexham) flying ground and surrounding farms in the mid-1930s. In February 1936 No. 610 ('County of Chester') Squadron formed as a light bomber unit at Hooton Park, taking possession of one of the Belfast aeroplane sheds to house its Avro Tutors and Hawker Harts when flying commenced in the May. Hooton Park was again taken over officially by the RAF in October 1939. Assorted, and completely unsuited, units took over convoy and anti-submarine duties in Liverpool Bay and along the North Wales coast, inevitably contributing to the slowly escalating list of crashes and fatalities in Flintshire and Denbighshire.

It is often stated that Tern Hill was selected as a First World War training airfield primarily on the recommendation of Major Atcherly, Chief Constable of Shropshire who, as noted in an earlier chapter, made a forced-landing at nearby Stoke Grange in one of Shrewsbury Flower Show's errant balloons on 22 August 1906. This is hardly creditable, but it makes a nice story — and other airfields have even more haphazard origins! Suffice it to say that ten years later land was requisitioned, fields levelled, hedges cleared, and a small hutted camp with canvas hangars set up.

The semi-official history of Tern Hill, compiled in 1975 when the RAF relinquished control to the Army, contains a vivid description of the early 'pioneer' days:

'Life during the early days of Tern Hill's existence was hard. The huts were small and lacked any kind of furniture, including beds; there was no electricity and the only heating was from small coke stoves, one in each hut. There were no proper kitchens or permanent kitchen staff; the cooking was done by anyone who happened to be fit enough to cut through the frozen meat which regularly arrived on the station. Each man was given three blankets and told to make his own sleeping arrangements. The morning had little more to offer. Due to the shortage of dishes and the absence of butter or margarine, breakfast consisted of stew eaten from tin bowls and slices of bread dipped into hot fat. The meal was finished with generous quantities of strong hot tea, the remains of which was saved for shaving water... !

'The working day during those early months involved most people in the preparation of both the buildings and the airfield for the arrival of the main

Flying Units, Nos. 34 and 43 Reserve Squadrons and Nos.30 and 33 Training Squadrons, mainly equipped with the rotary engined Bristol two-seaters and Avro 504s. Apart from one major setback on 6 October 1916 when two hangars were almost destroyed in a gale and several people were injured, work progressed smoothly and by the end of the year the station was operational. On 1 December 1916 the first supply of electricity came to Tern Hill. This was generated on the station by a single cylinder horizontal paraffin engine, with a similar engine in reserve. Shortly afterwards baths, ablutions, showers and a permanent hot water system became available to all. Life at Tern Hill now became very pleasant and continued in this way throughout 1917 and 1918, with the number of personnel gradually increasing to about 1,000 and the hangars housing nearly 50 aircraft'

Such descriptions will no doubt strike a responsive chord amongst Second World War airmen who had the task of opening up Shropshire airfields in the middle of the harsh winters of 1940/41 and 1941/42. Flying at Tern Hill began in December 1916 with the arrival of Nos. 34 and 43 (Reserve) Squadrons RFC with their Avro 504s and Sopwith Camels, soon augmented by 30 and 33 Squadrons, Australian Flying Corps. No. 4 TDS was established 1 September 1917 but moved almost immediately (19 September) to Hooton Park. A regular squadron, No. 95, formatted at Tern Hill on 8 October, but its tenure was even shorter than that of 90 Squadron at Shawbury! Three weeks later the cadre was packed off to Shotwick (Sealand) where for seven months it would flirt with oblivion before the axe finally fell.

With further expansion and changes in training policies Tern Hill became 13 TDS with three new squadrons specialising in the training of bomber crews. 132 Squadron was formed on 1 March 1918 as a day bomber unit, but commenced training on Sopwith Camels. It was disbanded on 23 December 1918, failing to become operational, although in September its cadre had moved to Castle Bromwich to replace 115 Squadron, already on the continent, to become part of the Independent Force being sent to France in the last two months of the war for the strategic bombing of German industrial targets. However, the Armistice intervened.

At the same time 133 and 134 Squadrons were also embodied at Tern Hill as night bomber squadrons and were intended to receive the Handley Page 0/400 twin engined bombers. While some sources doubt

An Instructor and Flight Cadet at 13TDS Tern Hill in front of a Maurice Farman MF.7 'Shorthorn'. [SRRC]

A Handley Page 0/400 with Sunbeam Maaori engines, possibly C9767, at Tern Hill, 1918. [Raymond Davies]

if any were delivered before the squadrons were disbanded, photographic evidence, although subject to varying interpretations, shows that at least one was received. 0/400 serial C9767 was, or had been, on the books of No. 1 (Communications) Squadron, formed at Hendon in December 1918 specifically to ferry passengers and Government officials to and from the Versailles conference in Paris. The unit moved to Kenley in April 1919, remaining there until the September. It would be re-designated 24 Squadron in February 1920. The presence of C9767 at Tern Hill for a photo-call late in 1919 may indicate either a visit by officials to the aerodrome, possibly to discuss its closure, or that it had been taken on charge following the dispersal of 0/400s from the communications squadron and No. 2 (Hendon) Aircraft Acceptance Park, which latter closed during the summer of 1919. Official records show at least three other 0/400s being taken on strength of 13 TDS —C9637 delivered from 2 AAP on 25 April 1918, C9639 on 1 June, and C9737 on 25 April 1918. These arrivals notwithstanding FE2bs were mainly used for training until both squadrons were disbanded on 4 July 1918 to provide reinforcements for active service units. 133 Squadron started to reform on 28 October

An Armstrong Whitworth FK.8, B4176, rebuilt from salvage and spares at Tern Hill. [RAF Museum]

1918, but the end of the war two weeks later put a stop to that.

The four Handley Page 0/400s languished on at Tern Hill until destroyed in a hangar fire on 9 March 1919. On the rare occasions when they took to the air they would visit Shawbury, a convenient limit and turning-round point to cross-country navigation exercises. Imperial War Museum photographs show piles of wreckage in fields 'near Shawbury', allegedly the demise of two of Tern Hill's 0/400s. The one, unidentified and attributed to 29 July 1917, is an impossibility as the first 0/400 did not reach the RAF until April 1918. The second, from the number visible on its starboard fin, was D4593, 'crashed near Shawbury on 22 August 1918'. The details are incorrect as records show D4593 being delivered to 14 AAP, Castle Bromwich, in October 1918 and crashing on 29 October at Maxstoke between Birmingham and Coventry. Fabric was stripped off the port wing in flight and the aircraft dived into the ground with its two Eagle VIII engines running flat out. Its seven crew — normal complement three or four men were killed instantly — Lt. R. E. Macbeth, Lt. F. J. Bravey, AM1 J. A. May, AM3 C. Offord, AM2 A. J. Winrow, AM2 H. A. Simmond, and AM3 G. Greenland. This suggests that the aircraft was on an air test, possibly pre-delivery to Tern Hill.

The *Shrewsbury Chronicle* for August and September 1918 carries some interesting pen pictures of life in the RAF, which junior Service had come into being on 1 April 1918 with the amalgamation of the RFC and RNAS. News of such an event was slow to reach rural back areas, hence the commendable initiative shown by the editor of the *Chronicle* in sending a reporter out to Tern Hill. The resulting 'purple' copy reads like the 'blurb' in a present-day holiday brochure. But such polemic follows awkwardly upon, and were therefore possibly commissioned as an antidote to, the spate of rather gruesome coroner's inquests in the wake of Tern Hill-based accidents that appeared in the same newspaper during June–August.

'Flying is the finest thing in the world for a boy or young man with a sporting spirit who wants to get on… ' — so warbles the cynical hack, trying hard to rekindle some of the patriotic fervour that was wearing a bit thin after four years of war and terrible wastage of human resources. One senses he himself remained to be convinced as he sought to assure his readers that promotion in the RAF 'was by merit alone', and that substantial private means were no longer essential to support officer status or to provide him with a flying machine! Although the war was drawing to a close, there was still a great demand for aircrew 'for aeroplanes, seaplanes, kite-balloons, and airships, rigid and non-rigid'.

If this were not enticing enough, he takes off his rose-tinted spectacles and gives it straight from the shoulder: '. . . also it must be borne in mind that he [the airman] always has a comfortable bed to sleep in, and he misses all the gas, the liquid fire, heavy bombardments and wet and muddy trenches that fall to the lot of his less fortunate companions-in-arms on the ground. And supposing he had the bad luck to be taken prisoner, it is a well-known fact that the Germans treat the airmen prisoners better than they do the rest'.

In October, in something of a contrived scenario, the privileged readers of the *Chronicle* were given an insight into the Officers Mess at Tern Hill: 'If a colonel of Crimean days were to stroll into the present-day RAF mess he would probably think that he had walked through 'Alice's looking-glass'. A cheerful youth welcomes the veteran and hospitably offers him a drink. From the old colonel's experience this youth should be a second-lieutenant. But on the cuffs of his well worn uniform the colonel notes with amazement the badges of a major's rank. On his breast, moreover, in addition to a pilot's 'wings' are sewn the ribbons of the D.S.O. and MC. This young airman's features wear the expression of mingled responsibility and dare-devil recklessness. Only gradually does the veteran realise he is speaking to the Commander of the Station, the owner of a name famous for masterly organisation and heroic gallantry in fight.

'Nearby lounges a grey-haired, somewhat portly man, who, despite his uniform, wears the air of a prosperous city merchant. His sleeves carry but one band, so he is a lieutenant. This is the Equipment Officer and a past master in the art of producing all the spare parts needed to keep aeroplanes constantly in the air and to keep a squadron happy. The youthful CO evidently holds him in high esteem and he seems on the best of terms with the youths around him. A few boyish looking captains are dotted about the ante-room, all with the look on their faces a schoolmaster likes to see on the faces of his prefects. These are Flight Commanders and all are accustomed to responsibility such as does not often fall in peace time to the lot of men double their age.

'Another middle-aged, well set-up officer next catches the veteran's eyes. He wears the uniform as if born in it and on his breast is the white-edged crimson ribbon of the Long Service and Good Conduct Medal. He is the Quartermaster, now performing in the RAF the duties which for years he has performed for a regular battalion in the Army. Harder to place is a second-lieutenant without badges or bands of any sort, with bowed shoulders and keen intellectual face. He proves to be the Mechanical Transport Officer and what he does not know about economy in petrol, repairs to engines and organisation of journeys, is not worth knowing.

The rest of this mess is made up of some youthful instructors and yet more boyish pupils, who listen modestly to the stories of the 'old hands' about battles with enemy. A surprising collection of men they appear to the old colonel, but are full of the right spirit, all animated by esprit de corps, and are engaged in building up the traditions which are to be handed down to future generations of the RAF. After being in their company for an hour or so, the Crimean veteran would probably admit that he found no falling off here in the quality of the youngsters who held the King's commission'.

All stirring stuff. What sprig of the middle-class could not resist the urge to make for the nearest RAF recruiting depôt, while the lower orders could put down their paper with the comfortable feeling that in such hands all was well with the world?

But it was a case of 'No names, no pack drill!' And probably deliberately so since, on Tuesday, 30 July 1918, Tern Hill had lost its previous Commanding Officer under particularly horrifying circumstances. Shortly before mid-day, over Spoonley, near Market Drayton, his plane was seen at about 6,000ft manoeuvring as if to loop the loop. It suddenly slipped sideways and an object was seen to fall out. The aircraft turned over onto its back and crashed into the ground at full throttle, disintegrating completely. The body of the RAF mechanic, 2nd-class Private Fred Lythgoe, 18, from Atherton, near Manchester, was found twelve yards from the plane. The remains of Major C. E. Brisley, 32, a war veteran, were found in a field 200 yards away. His watch had stopped at 11. 48am. The machine, a new Avro 504K (serial D6361), had been flight tested the previous evening. Young Lythgoe was due to become an NCO and his CO was putting him through his aptitude tests. Both were strapped in, but the pilot obviously not tight enough and had slipped his straps. It emerged at the inquest that many pilots dispensed with safety harness; indeed, belts were still not fitted to certain types of aircraft. Major Brisley lies buried in Market Drayton cemetery.

By strange coincidence Major Brisley himself had been scheduled later that day to attend a coroner's inquest on two Shawbury pilots who had been killed whilst flying from his aerodrome. On Sunday, 28 July, Flight Cadet C. E. Brown, 18, had made a normal solo take-off from Tern Hill, but at 300ft his machine (A BE2e serial A1391) went into a spin, hit the ground and burst into flames. Verdict: 'He lost his head and failed to shut off engine when the machine descended'. On the Monday morning, 2/Lt. J. A Freeman, 28, with 35 hours solo already logged, made a straight-forward take-off in an RE8 (serial C2397), but then turned right over and fell wing over wing to earth. The petrol tank blew up in a terrific blaze and a column of smoke seen for miles. The unfortunate pilot was killed on impact and his body badly burnt. Small wonder that the subsequent PR visit to the Officers' Mess turned out to

Airco DH.4, B6416, christened 'The Straffer' of 10TS Shawbury comes to grief in Ludlow Park, 13 March 1918, whist taking part in a War Bonds promotion drive. [RAF Museum]

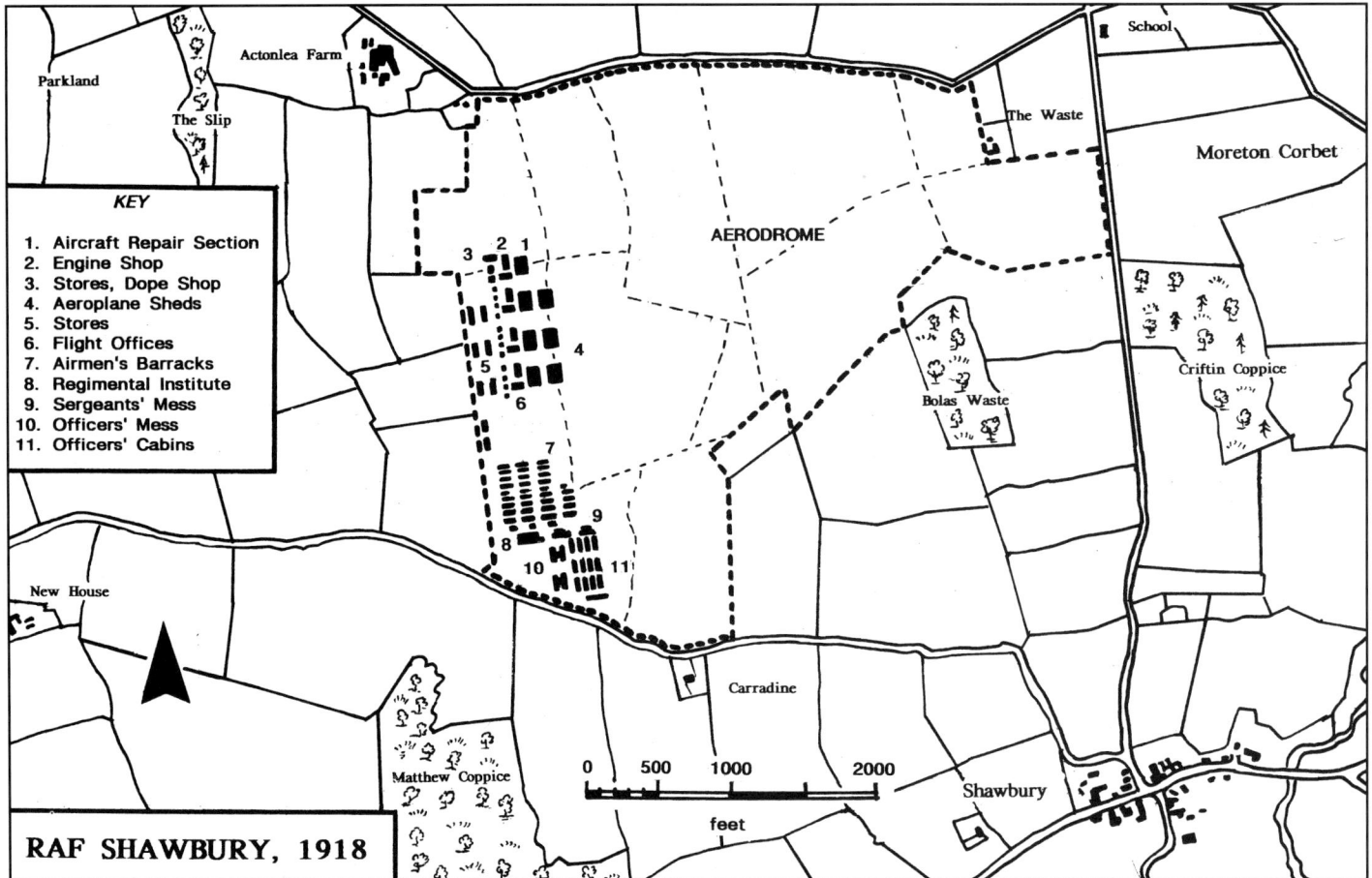

KEY

1. Aircraft Repair Section
2. Engine Shop
3. Stores, Dope Shop
4. Aeroplane Sheds
5. Stores
6. Flight Offices
7. Airmen's Barracks
8. Regimental Institute
9. Sergeants' Mess
10. Officers' Mess
11. Officers' Cabins

RAF SHAWBURY, 1918

be a rather subdued, almost anonymous affair!

In the wake of the Armistice there would follow a short period of indecision, whether or not to retain Tern Hill as a permanent RAF station. A couple of squadron cadres arrived during 1919 for disbandment and demobilisation. The airfield was closed in 1920 and sold off two years later. In 1934, as part of the 'Expansion' programme Tern Hill was re-surveyed, approved, and land re-requisitioned for a training establishment. Building work began in 1935 and the station commissioned on 1 January 1936 with 10 FTS in residence, flying training beginning on 3 February 1936. From 1 June 1937 it also housed No. 4 Aircraft Storage Unit, later 24 MU.

Tern Hill will loom large in our later narrative on two counts. Before 9 Group Fighter Command became operational, and 96 Squadron was installed at RAF Cranage, and before the opening of operational airfields at Wrexham and High Ercall, Tern Hill also operated as an *ad hoc* fighter station, offering stabling on a daily and nightly basis to such detachments as could be spared by 12 Group, such as 29 and 611 Squadrons. For six vital months in 1940 Tern Hill would offer the only air defence on the southern approaches to Wrexham and Merseyside. Secondly, in November 1940, when 5 SFTS with its Miles Masters moved from Sealand to Tern Hill, it would simply be to shift slightly southwards into the Berwyns and the Shropshire hills, the epicentre of an already unacceptably high accident rate associated with landings and take-offs, collisions, low flying and aerobatics.

Fourteen miles to the south of Tern Hill the attention of engineers had focussed, simultaneously on the flat area adjacent to the old Roman Road (B5063) between Moreton Corbet and Shawbury. Free of obstruction and mainly in the ownership of one person, over 300 acres were requisitioned in 1917 and two 1,000-yard landing strips laid out. Three training squadrons took up residence during June and, after some unit exchanges with Tern Hill, commenced flying with a motley

collection of aircraft on 1 September 1917. An Aircraft Repair Section or Squadron was established at the same time, working closely with Hooton Park ARS, receiving and testing all new aircraft destined for Shawbury and Tern Hill as well as being responsible for repair, maintenance and salvage. An attempt to establish and work up an operational fighter unit, 90 Squadron, quickly foundered. Embodied at Shawbury on 8 October 1917, pressure on space forced its move to Shotwick (Sealand) on 5 December.

The five training squadrons at Shawbury and Tern Hill came under the aegis of No. 29 (Training) Wing. Originally three of them were Australian with training methods differing slightly from those employed by their British counterparts. No. 29 TS (Aust.) at Shawbury accepted pupils who had successfully passed the initial training course for prospective pilots and observers at No. 1 School of Military Aeronautics at Reading (Coley Park). After completing an 8-week 'intermediate' course at Shawbury the Australians moved on to 30 TS (Aust.) at Tern Hill for advanced training. On the other hand, British pilots in Nos. 10 and 17 Training Squadrons received 'all through' instruction and were posted directly to operational squadrons.

The first Americans arrived for training at Shawbury in November 1917, but to them cannot be attributed credit for all breaches of King's Regulations (or 'wild spirits') such as taking girl friends for a 'flip', flying under Shrewsbury's bridges, or beating up open air concerts, garden parties and cricket matches. The *Shrewsbury Chronicle* could well bemoan the fact, that on Monday evening, 16 July 1917, the concert in the Quarry was spoilt by the manoeuvres of 'a powerful bi-plane' that for thirty minutes looped the loop 'time out of number' and 'flew upside down and indulged in a variety of apparently reckless antics at high and low altitudes'. It is doubtful if the pilot was even aware of the cultural event taking place 2,000ft below him; if he were, who could blame him for showing off a little?

Later in July, on the last night of Shrewsbury School's 'bumping

RFC/RAF STATIONS IN NORTH WALES AND THE BORDER, 1914–19

UNIT	IN	FROM	OUT	TO
BANGOR/ABER (Glan-y-môr-isaf)				
244 Sqn.	15.8.18	Formed from 521 522, 530 SDFs	22.1.19	Disbanded
521 SDF	15.8.18	Llangefni	22.1.19	Disbanded
522 SDF	15.8.18	Llangefni	22.1.19	Disbanded
530 SDF	15.8.18	Formed	18.10.18	Tallaght (Dublin)
530 SDF	11.11.18	Tallaght	22.1.19	Disbanded
HOOTON PARK				
4 TDS	19.9.17	Tern Hill	14.3.19	Became 4 TS/TSch
43 Aero Sqn. USAS	-.8.18	South Carlton	-.10.18	France
HQ 37 Training Wing	3.11.18	Ledsham, Little Sutton	9.4.19	Disbanded
117 Sqn.	8.3.19	Wyton	18.3.19	Tallaght (Dublin)
4 TS/TSch	14.3.19	From 4 TDS	31.5.19	Disbanded
LLANGEFNI/ANGLESEY				
RNAS (Airships)	26.9.15	Commissioned	24.1.19	Care and maintenance
521 SDF	6.6.18	Formed in 255 Sqn.	15.8.18	Aber (Bangor)
522 SDF	6.6.18	Formed in 255 Sqn.	15.8.18	Aber (Bangor)
MONKMOOR				
OSRAP	19.10.18	Formed	2.5.19	Disbanded
14 AAP Sub-depôt	-.9.18	Established	31.10.19	Relinquished
SHAWBURY				
67 TS	11.6.17	Castle Bromwich	31.3.18	Shotwick
29 TS(Aust)	15.6.17	Formed	14.1.18	Became 5 TS AFC
30 TS(Aust)	15.6.17	Formed	30.6.17	Tern Hill
10 TS	30.6.17	Tern Hill	7.4.18	Lilbourne
HQ 29 Wing	5.8.17	Tern Hill	9.4.19	Disbanded
Aircraft Repair Section	1.9.17	Formed	31.10.18	Disbanded
90 Sqn.	8.10.17	Formed	4.12.17	Shotwick
5 TS AFC	14.1.18	From 29 TS(Aust.)	2.4.18	Minchinhampton
131 Sqn.	1.3.18	Formed	17.8.18	Disbanded
9 TDS	15.3.18	Formed	25.4.19	Disbanded
137 Sqn.	1.4.18	Formed	4.7.18	Disbanded
SHOTWICK				
90 Sqn.	5.12.17	Shawbury	5.7.18	Brockworth
95 Sqn.	30.10.17	Tern Hill	4.7.18	Disbanded
96 Sqn.	30.10.17	South Carlton	4.7.18	Disbanded (?into 51 TDS)
55 TS	-.7.18	Narborough	15.7.18	Disbanded into 51 TDS
67 TS	31.3.18	Shawbury	15.7.18	Disbanded into 51 TDS
51 TDS	15.7.18	From 55/67 TS	14.3.19	Became 51 TS
51 TS	14.3.19	From 51 TDS	31.5.19	Became 4 TSch
27 Sqn. cadre	18.3.19	Buvay	22.1.20	Disbanded
98 Sqn.	21.3.19	Alquines (France)	24.6.19	Disbanded
103 Sqn.	26.3.19	Maisoncelle (France)	1.10.19	Disbanded
4 TSch	31.5.19	From 51 TS	26.4.20	Became 5 FTS
HQ 13 Training Group	26.1.19	Upton	18.10.19	Disbanded into 3 Tr/Gp
HQ 10 TW	18.10.19	Reformed	7.2.20	Disbanded
55 Sqn. cadre	1.1.20	Renfrew	22.1.20	Disbanded
5 FTS	26.4.20	From 4 TSch		

TERN HILL

34 Res Sqn.	13.11.16	Castle Bromwich	1.6.17	Became 34 TS
43 Res Sqn.	13.11.16	Castle Bromwich	1.6.17	Became 43 TS
33 Res Sqn. (Aust.)	-.12.16	Formed	-.5.17	Cirencester
63 Res Sqn.	28.3.17	Formed	1.6.17	Became 63 TS
HQ 29 Training Wing	1.6.17	25 Tr. Wing	5.8.17	Shawbury
10 TS	1.6.17	Joyce Green	30.6.17	Shawbury
34 TS	1.6.17	From 34 RS	18.3.18	Chattis Hill (Hants)
43 TS	1.6.17	From 43 RS	20.3.18	Chattis Hill
63 TS	1.6.17	From 63 RS	1.6.17	Joyce Green
30 TS (Aust.)	30.6.17	Shawbury	14.1.18	Became 6 TS AFC
4 TDS	1.9.17	Formed	19.9.17	Hooton Park
95 Sqn.	1.10.17	Formed	29.10.17	Shotwick
6 TS AFC (Aust.)	14.1.18	From 30 TS	25.2.18	Minchinhampton
13 TDS	1.3.18	Formed	14.3.19	Became 13 TSch
132 Sqn.	1.3.18	Formed	19.8.18	Castle Bromwich
133 Sqn.	1.3.18	Formed	4.7.18	Disbanded
134 Sqn.	1.3.18	Formed	4.7.18	Disbanded
13 TSch	14.3.19	From 13 TDS	-.3.20	Disbanded

races' on the Severn, word had somehow got around that 'an aerial visitation' was imminent. Hundreds of people lined both banks either side of the Kingsland Bridge. Earlier in the afternoon two aircraft had landed on the School's cricket field, the one piloted by Old Salopian Lt. (more correctly Capt) Ellert Webster Forbes, MC, Royal Warwickshire Regiment and RAF. He had had several narrow escapes as an observer in FE2bs with 20 Squadron in France and had gained his decoration on 16 May 1916 when, although wounded, he had successfully brought down his aircraft (serial 6359) after his pilot had been killed in combat over Ypres. He could be forgiven for writing it off! In July 1917, after hospitalisation, he was completing his pilot training at Shawbury. He was joined on the cricket field by a Capt. Collet, who, as *The Salopian* proudly informed parents 'on previous occasions has given us such thrilling displays', continuing '. . . but on Saturday night he eclipsed all previous performances and electrified the waiting masses by flying not once but some nine or ten times under Kingsland Bridge', thus possibly establishing the precedent followed by many later Shawbury pupils after successfully 'passing out'.

In step with the radical changes at Tern Hill, Shawbury became No. 9 Training Depôt Station on 1 March 1918, but it was several weeks before the old training squadrons were disbanded and shipped out, their places being taken by 131 Squadron (1 March 1918) and 137 Squadron (1 April 1918). These two squadrons were respectively intended as day and night bomber units equipped with DH9s but in practice had to make do with a rag bag of DH4s, DH6s, RE.8s, Avro 504J/Ks, and BE2s. The two squadrons never became operational, and were disbanded on 17 August and 4 July 1918. However, the TDS continued to send pilots to active service squadrons until after the Armistice, when it, too, was disbanded (25 April 1919). Between 1 July–31 December the aerodrome functioned as a subsidiary surplus aircraft storage facility to 14 AAP, Castle Bromwich. Shawbury would remain open as a transit camp and demobilisation centre until January 1920, finally closing in May of that year.

The interlude was only a brief one. In 1937, under the 'Expansion Programme', work began on a brand new permanent station, several times larger than the original. It would house No. 27 Maintenance Unit (1 February 1938) and No. 11 Flying Training School from RAF Wittering, Northamptonshire (14 May 1938), which latter airfield was now required for operational use. Thus would begin the intricate relationship of RAF Shawbury with almost every other airfield in North Wales, Cheshire and Shropshire, but in particular with Calveley, Cranage, and Wrexham.

Being a training establishment accidents and fatalities were bound to occur out of Shawbury, one of the earliest, as already noted, being 2/Lt. A. C. Tallent, whose forced landing on 2 October 1917 at Overton (by Malpas) gave south Cheshire its first taste of the air war. He was quickly followed on 13 October 1917, by 2/Lt. Harold James Cryer, 10 TS, whose Sopwith Camel (B5164) nosed-dived in from 1,000 ft., killing him instantly. In St. Mary's churchyard, Shawbury, some fifty graves of airmen of both wars bear witness to the price of striving after perfection.

Along the Dee

There are in Hawarden's two cemeteries some three dozen or so military graves relating to Second World War units at RAF Hawarden and its satellite at Poulton north of Wrexham. There are also fifteen connected with RAF Sealand covering the period 1922 onwards, the year in which No. 5 Flying Training School really got into its stride and when at least eighteen aircraft fell from the skies and seven would-be pilots tragically had their flying careers cut short. On the Cheshire side of the Dee estuary, in the tiny churchyard of St. Michael's, Shotwick, a further eight graves underscore the fact that, for the first decade of its existence, until 1924, Sealand was known officially as RAF Shotwick, after the custom of naming airfields from the nearest village. There were nine graves originally, but that of 19-year old Lt. Hugh Robert Futtr, 67 Training Squadron, formerly of the 12th South African Infantry Regiment, killed in a flying accident on 16 June 1918, was exhumed shortly after the war.

Today Shotwick lies off the beaten track, properly accessible only off the A550. But until rights of way were finally extinguished in 1923 the Green Lane leading from Queensferry (King's Ferry) and representing the ancient track across the Dee marshes, gave access from the south and off the aerodrome. It was the encroaching settlement of Sealand Garden Suburb, begun in 1910 for the employees of John Summers & Co.'s steelworks, that, in brick and mortar terms, eventually tied the multi-site airfield to industrial Flintshire at Queensferry in preference to rural Shotwick. Add to this some apocryphal stories of confusion with RAF Scopwick in Lincolnshire… and time was evidently ripe for a change of name. However, it would appear that Scopwick changed its name first, becoming RAF Digby in July 1920!

What prompted the Air Board to build not one, but originally two,

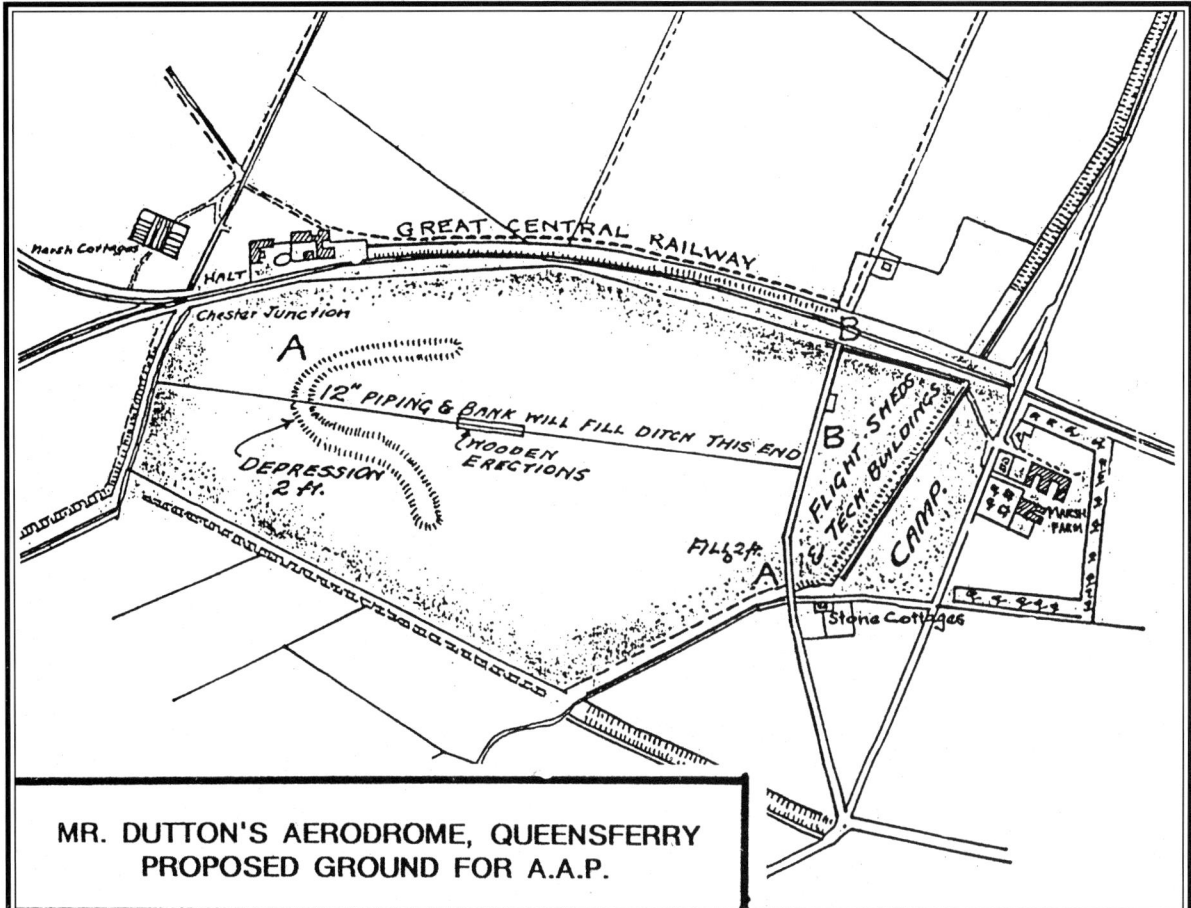

MR. DUTTON'S AERODROME, QUEENSFERRY
PROPOSED GROUND FOR A.A.P.

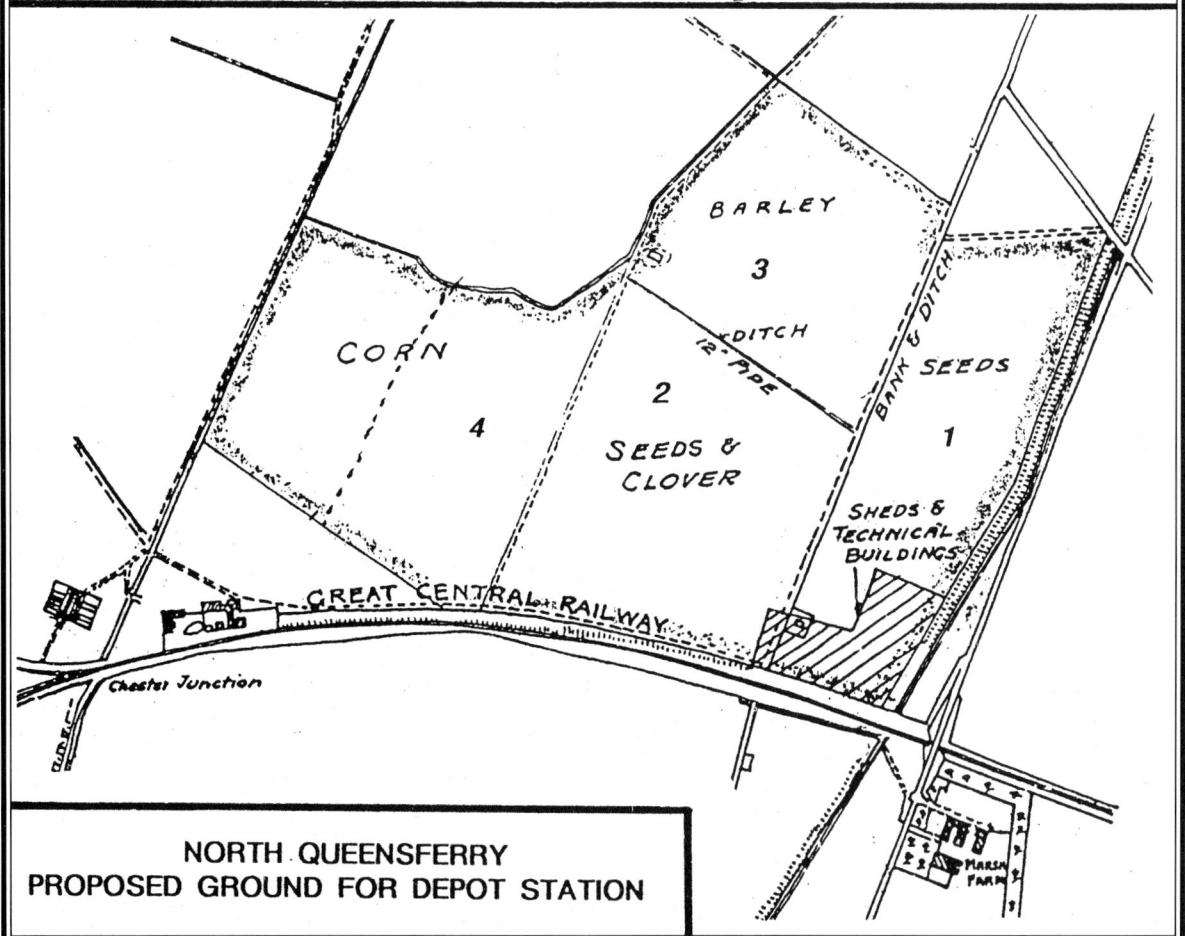

NORTH QUEENSFERRY
PROPOSED GROUND FOR DEPOT STATION

This aerial view c.1930 shows RAF Sealand very much as it was at the end of 1918. A key map is supplied below. Some post-1927 remodelling has taken place following the aerodrome's retention in to peace time. [FRO]

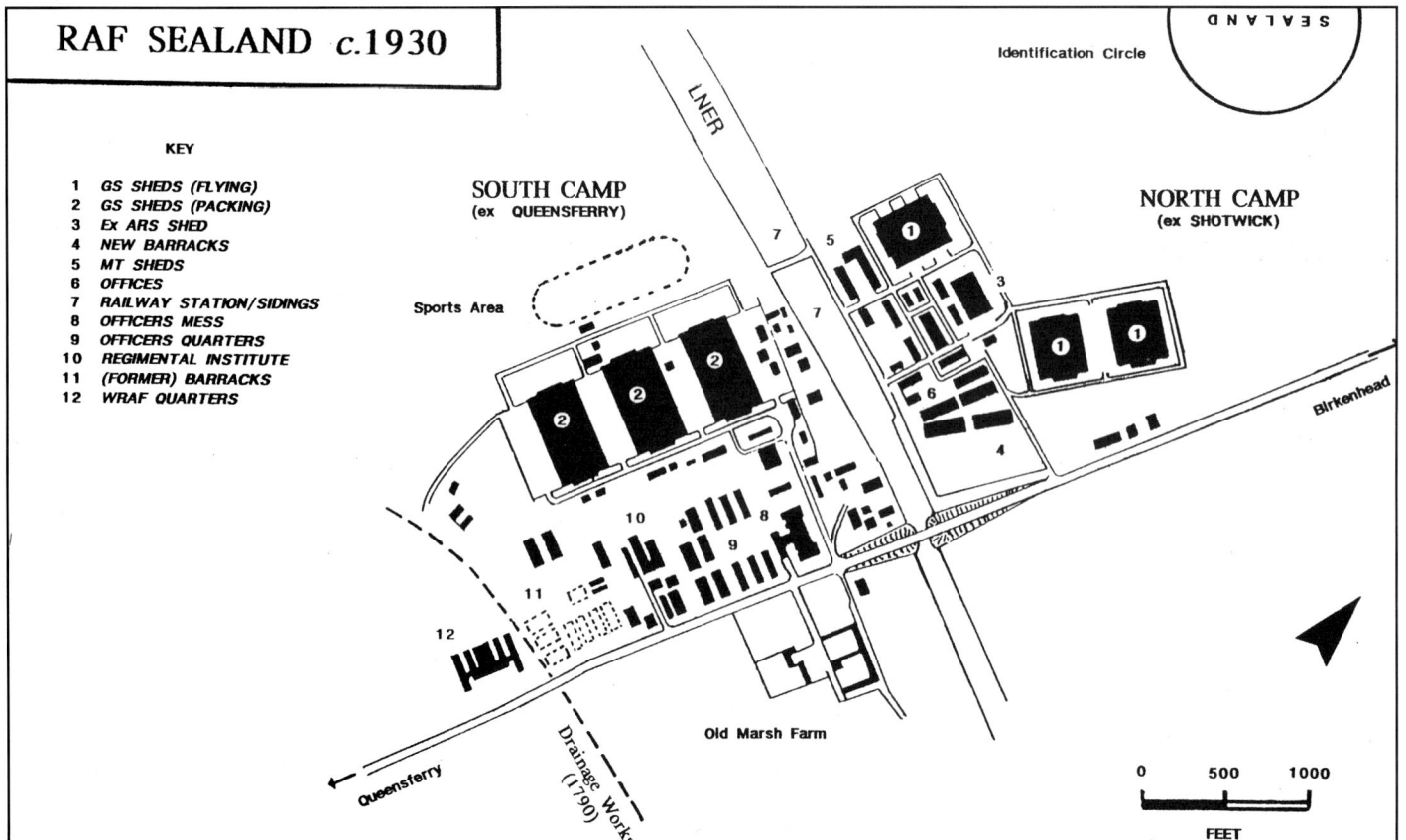

RAF SEALAND c.1930

KEY

1 GS SHEDS (FLYING)
2 GS SHEDS (PACKING)
3 Ex ARS SHED
4 NEW BARRACKS
5 MT SHEDS
6 OFFICES
7 RAILWAY STATION/SIDINGS
8 OFFICERS MESS
9 OFFICERS QUARTERS
10 REGIMENTAL INSTITUTE
11 (FORMER) BARRACKS
12 WRAF QUARTERS

SOUTH CAMP
(ex QUEENSFERRY)

NORTH CAMP
(ex SHOTWICK)

Sports Area

Identification Circle

SEALAND

LNER

Birkenhead

Old Marsh Farm

Queensferry

Drainage Works (1790)

0   500   1000

FEET

airfields at Sealand is anybody's guess. It would appear that logistical factors, such as the proximity to two railway lines and the port of Liverpool, were permitted to obscure minor physical disadvantages like the land being only 15–18ft above sea level, with a high water-table, liable to turn into a morass in wet weather, and with a tendency for the alluvial sands and loams to 'blow' over protracted dry spells.

Since 1754, as its name implies, Sealand township had been undergoing piecemeal reclamation from the sandbanks, marshes and saltings of the Dee estuary following the canalisation of the river seventeen years earlier. Indeed, the great breach in the 'Broken Bank', the last dyke to be completed in 1877 and almost immediately ruptured by north westerly gales, would only be finally closed in 1916. T. M. Dutton's Queensferry flying ground, which would be requisitioned as an Air Acceptance Park (otherwise South Shotwick), was located on land between the 1763 and 1833 dykes and formed part of (Old) Marsh Farm. North Shotwick aerodrome, home to the First World War training squadrons, an Aircraft Repair Section, and later 5 FTS, would be built on land reclaimed in 1857, but which until the 1890s was useable only as rough sheep grazings and a series of thirteen 1,000 yds. Volunteer target ranges. By 1901 fields had been ditched and hedged and the area largely brought under cultivation as part of Howard's or New Marsh Farm. The firing ranges moved west of the dyke, an area that would be used in the Second World War as a bombing range and Puddington 'Q-site', a Q/F decoy for RAF Sealand.

Drainage was a shared responsibility, each tenant/owner scouring and maintaining water courses crossing his land. When aerodrome contractors moved onto the site in September 1917 the very delicate balance was wrecked, 12-inch pipes being no substitute for 12ft open drains. Former creek lines or depression were in-filled, but as made-up ground would always hold water they produced hazards which contributed to the demise of many a training aircraft. Wet winters, such as that of 1921 brought flying to a standstill for long periods, despite the constitution of a Sealand Drainage District in 1920. The winter of 1924/25 was particularly bad, but it was 2 April 1925 before things had dried out sufficiently for 'remedial' works to be set in hand. However, the problem of flooding was never satisfactorily solved and would have particularly serious repercussions during the Second World War when RAF Sealand was forced to farm out flying instruction to 'drier' airfields (relatively) at Hawarden (a case of 'out of the frying pan …!), Speke, Hooton, Cranage, even Ringway, and RLGs at Wrexham and Little Sutton.

An account of the visit to Sealand of Major Donald Clappen, Directorate of Aircraft Depôts and Acceptance Parks, survives, although written some sixty years after the event. The tenurial history of the area suggests that he actually visited and recommended a more suitable site further east towards Burton Point, land which had been reclaimed and farmed for over 125 years and where drainage problems were not so acute. For whatever reason his suggestion was not followed up, the Air Ministry, perhaps attracted by Dutton's pre-existing airfield and better rail and road access, opting for the sites nearer Queensferry. Since June 1906 George Ledson had been farming both Old and New Marsh Farms on both sides of the railway. On 5 September 1917 he received formal notice of intention to requisition some 224 acres under the Defence of the Realm Act, 1914. Major Thompson, a DA&AAP field officer, inspected the site — including Dutton's aerodrome that very same day. Within a week, before Mr. Ledson could appeal, the contractors were in and land under clover, roots and manured ready for grain crops disappeared under a RFC landing ground. Thus began the long drawn-out battle for compensation that was still going on in January 1926.

The aerodrome north of the railway ('North Shotwick') was opened for flying training in October 1917. That south of the railway ('Queensferry or 'South Shotwick') was intended as an Aircraft Acceptance Park, but was possibly only fully operational as such a few

months before the end of hostilities. The Aircraft Repair Section (ARS) under the command of Capt. Thomas Murthwaite Dutton was up and running by November 1917, housed in the half or single Belfast hangar at the eastern end of North Shotwick. In addition to the receipt and issue of aircraft to the training squadrons the ARS was also responsible for general maintenance of aircraft from the AAP across the railway track and salvaging those that came to grief in Flintshire, Denbighshire and Shropshire. It worked in close conjunction with the ARS, hence its official designation '37 Wing Half ARS'!

It is difficult to assign an official date from which AAP South Shotwick was up and running because latterly pressures were such that aerodromes and support installations were regularly occupied before completion. Additionally Aircraft Movement Cards and unit monthly/quarterly inventories and returns have not survived in any quantity. From the few that do exist one must assume that South Shotwick was up and running by at least August 1918 if not earlier. Some eighteen Norman Thompson NT.2B 'pusher' training flying boats are noted as taken on charge between 4 September–23 November 1918, eight as new (N2266, N2405–11), delivered straight from the makers at Middleton-on-Sea, and the remainder (N2561, N2563–67, N2570–73), ex-210 TDS, Calshot, from storage at AAP Eastleigh (Hampshire). Like South Shotwick the latter was unnumbered and as of July 1918 only half completed but in the process of being taken over by the United States Navy and therefore needing to transfer its RAF holdings elsewhere. South Shotwick was as strategically placed for Liverpool as Eastleigh was for Southampton. While most of the NT.2Bs were still on South Shotwick's books the week ending 30 January 1919, several had been exported to Peru, Norway, and for exhibition purposes to the USA. AAP South Shotwick's main role would therefore seem to have been moth-balling new machines surplus to operational requirements from contracts that had not been cancelled, and storing their share of the huge glut of redundant aircraft that materialised once hostilities had ceased.

In addition to its training squadrons North Shotwick would receive a couple of regular squadrons, Nos. 90 and 95 from Tern Hill and Shawbury. On 30 October a third regular unit was posted in, No. 96 Squadron formed three weeks earlier at South Carlton, Lincs. Thus began this squadron's association with Cheshire and North Wales, links that would be renewed and strengthened at Cranage and Wrexham in the critical years 1940–43. Before 96 Squadron's Defiants and Beaufighters cluttered up the border landscape there was 96 Squadron's Sopwith Camel (C8305) which on Wednesday, 12 June 1918, stalled on take-off and spun in out of control from 200 ft., killing its pilot, 2/Lt. G. C. R. Hamilton. The wreckage was moved across the railway to the ARS where it was quickly repaired, ending its life with No. 1 Flying School, Turnberry.

But for the moment all three squadrons would lead something of a 'Cinderella' existence. Intended as fighter squadrons standardising on Sopwith Dolphins, they had to make do with a motley assemblage of Pups, Martynsides, BE2cs etc. To add insult to injury, 95 and 96 Squadrons were disbanded on 4 July 1918 and their pilots dispersed to active service units in France. The following day 90 Squadron was posted to Brockworth, Gloucestershire, where it lingered on another month before its personnel, too, were dispersed.

Shotwick thus reverted exclusively to its training role. 67 Training Squadron moved in from Shawbury on 31 March 1918. It arrived empty handed having bequeathed its aircraft to newly formed 131 Squadron on 15 March. 67 TS poodled about North Shotwick for some three months before it was finally disbanded into 51 TDS, sufficient time for it to write off at least six aircraft and killing three pilots. Similarly 55 TS came in from the cold. Formed in Castle Bromwich on 11 May 1915 it had led, since 12 December 1917, something of a peripatetic existence first at Lilbourne (Northamptonshire) and then at Narborough (Leicestershire), arriving at Shotwick early in July 1918.

A Norman Thompson NT.2b 'pusher' flying boat of the type stored at South Shotwick AAP as surplus to requirements or prior to export via Birkenhead docks. [Raymond Davies]

The exact day is not known, but it must have been before 12 July when one of its Sopwith Camels, D8127, crashed near Shotwick, F/Cdt. T. A. Emmerton becoming disorientated in cloud and spinning in out of control. Three days later 5 TS was also disbanded into 51 TDS.

North Shotwick became No. 51 Training Depôt Station on 15 July 1918 by the simple expedient of disbanding the training squadrons in name and taking over their aircraft. In practice they continued to exist. No. 51 TDS was planned to have three day fighter training squadrons, not only taking trainees *ab initio*, but also giving advanced training to potential flying instructors. The impact was immediate as 51 TDS Camels and SE5as started to fall out of the sky, at least seventeen over the next six months with eight pilots killed. The nine graves in Shotwick's tiny churchyard are but representative tokens of the considerably greater loss of life that occurred at the nearby aerodrome. They cover the period 25 April to 9 November 1918.

Amongst them is that of Lt. Herbert Westgarth Soulby who was killed in a flying accident in Avro 504K H2250 on 19 October 1918. No pupil under instruction this, but a holder of the French Croix de Guerre and veteran of the Western Front. He was blooded in October 1916 as an observer flying FE.2ds in a reconnaissance role with 20 Squadron, before training as as a pilot. He was posted to 70 Squadron, flying Camels, surviving several crashes, and shooting down an Aviatik C over Becelaere on 4 January 1918 and an Albatros D.V over Quesnoy on 9 March. Fate ordained that he should be posted to back-area Flintshire to die, as a visiting instructor up-dating staff on tactics and aerial strategy. The headstones also indicate that Saturday, 9 November 1918 was a black day for the unit, losing F/Cdt. Vernon Francis Gibson, aged 19, whose Camel (F1946) stalled on a turn and spun in, and 2/Lt. Francis Athol Hinton, a New Zealander, who had just completed his training and was about to start on a flying instructor's course. His aircraft., an Avro 504K F8719, suffered engine trouble, lost flying speed, stalled and crashed at Buerton, between Saighton and Aldford, Cheshire.

As at Hooton Park, a large number of North Shotwick's trainees hailed from the Dominions and the USA. The graves in Shotwick's churchyard comprise three Americans, two Canadians, one South African and a New Zealander. Ever since 1929 the grave of Lt. Leonard S. Morange, from Bronxville, New York, has had flowers placed on it on the anniversary of his death (11 August 1918 in an Avro 504K serial E1719) by courtesy of the Bronxville branches of the American Legion and Boy Scouts.

Nine days before the death of Lt. Morange, a colleague, Lt. W. D. Archer had been persuaded by Thomas Murthwaite Dutton to pen some thoughts of an American under training in the UK for *The Mancot Circular*, a sort of community newsletter of the time: 'I was

flying a Sopwith Scout, a two-seater perhaps better known as 'Spinning Jenny'. . . . ' He recalled waiting for the morning mists to clear off the estuary before taking off from Shotwick, '... the dart across the green flying field with the cold wind whistling through the struts, then climbing through the mist and low cloud to burst into the bright sunshine, practising loops and climbs. Then after a while drifting down the River Dee at altitude to Chester before diving down over the Dee meadows towards Wrexham, climbing again to avoid dark rain clouds on approaching the hills west of Wrexham.' Then the last leg of the navigation exercise, turning north-west, picking up the Dee again and following it back to base. 'After an hour or so I stepped out of the plane frozen stiff. . . then to the Mess for an English breakfast!'

Romantic ideals masking serious intent! At least Lt. Archer would see some of the action. On 9 October 1918 he was on a low level bombing offensive patrol, attacking the vast columns of retreating German troops. He was flying a SE5a (E4089) of 40 Squadron out of Bryas. At 16.45 hrs. over Petit Fontaine, west of Cambrai, his radiator and engine were hit by machine-gun fire from the ground. In making a forced landing E4089 ran into a shell hole and overturned, ripping off the undercarriage and injuring further its pilot who had already sustained a nasty bullet wound.

But veteran or green heads on young shoulders, the slightest lapse of concentration and the result was the same. Capt. J. H. Tudhope, a South African, was flight commander and 2i/c 51 TDS. On 4 February 1919 he was almost killed by the propeller of his Sopwith Camel (D8130). As it was he escaped with only slight injuries, torn flying rig, and dented ego. He should have known better, being a Western Front 'ace', flying Nieuport 24s and SE5as with 40 Squadron based at Bruay in the Pas de Calais. He gained the first of his ten victories in Nieuport 24 B3617 at 08.30 hrs. on 22 September 1917 when he downed an Albatros D.V east of Pont a Vendin. Similarly Lt. M. S. Pettit, ex-54 Squadron, who had been shot up during an offensive patrol on 27 March 1918 (in Camel C1667), found his posting as an instructor to Hooton Park more than hazardous, downright fatal in fact. On 11 August 1918 he was dead, dying of shock and multiple injuries on his way to a Chester hospital after a mid-air collision in which his aircraft had its tail ripped off by that of a pupil trainee going solo for the first time.

It was planned that 51 TDS should ultimately have an establishment of 36 Avro 504s and 36 Sopwith Dolphins. These were very slow in appearing and for eight months 51 TDS had to struggle along with half that number of miscellaneous aircraft. Indeed, the third unit or training squadron within the TDS was never formed. North Shotwick operated at two-thirds of planned strength before the drastic post-Armistice reduction in the size of the RAF removed the need for new pilots. Further run down was inevitable despite the TDS accepting and 'finishing off' partly trained pupils from other units that had been closed.

As noted in an earlier chapter, a decision to retain Shotwick as a permanent peace-time training establishment had possibly been taken in principle in early March 1919, hence the use of the base as a demobilisation aerodrome for units arriving from France with or without aircraft. First to arrive on 18 March 1919 from Bavay (Bavai) on the French/Belgian border was 27 Squadron, as a cadre only. HQ staff and a few personnel loitered on at Shotwick until 22 January 1920 when it was finally disbanded. Also disbanded on 22 January was 55

Squadron cadre which had only arrived at Shotwick on 1 January after kicking its heels at Renfrew for some eleven months. 98 Squadron, cadre only, arrived on 28 March 1919 from Alquines. It would disband on 24 June. Also on 28 March arrived 103 Squadron from Maisoncelle. Although a cadre it may have brought some of its DH9s with it. It was finally disbanded on 1 October 1919.

North Shotwick's training side was given a much needed boost with the arrival of the rump of 4 TDS from Hooton Park which aerodrome was being closed down. The exact date is uncertain, but was possibly around 23 March 1919 after 177 Squadron, the last RAF unit to use the aerodrome, had left Hooton Park en route for Tallaght and internal security duties in Ireland. One notes that at this period Shotwick's 'Irish connection' was also being strengthened with the movement in February/March 1919 of surplus SE5a fighters, some ex-85 Squadron currently languishing at Lapcombe Corner, Wiltshire, to Ireland and 11 (Irish) Group.

It seems that 4 TDS became 4 TS on transfer to Shotwick, possibly becoming the long awaited third unit or squadron of the planned establishment. Some records suggest that in practice 4 TS, as a re-designated 4 Training School, may have taken over the training role from 51 TDS, ultimately absorbing the latter. Whatever, the sudden drop in trainee numbers had its positive side — pupils now worked to a properly constructed integrated syllabus, the courses being lengthened first to twelve then to sixteen weeks.

Although the permanent acquisition of land and severance of public rights of way were not completed until 1923, the merger of the surviving elements of 51 TDS and 4 TS to form No. 5 Flying Training School took place in April 1920. Its first Commanding Officer was W/Cdr. Francis Know Haskins, a former RNAS pilot, who in October 1915 had flown anti-Zeppelin patrols out of St. Pol with 1 (Naval) Squadron and would command that squadron between December 1916-June 1917.

Despite the pre-war proselytizing of Gustav Hamel, Benny Hucks and others on the North Wales and Shropshire flying circuit, it was a long time before aeroplanes became common place in the Wrexham area. In the Parkey, Bowling Bank, Redwither and Ridley areas along the Dee the sighting of an aeroplane was still a topic of conversation in 'The Kiln', 'The Nag's Head' and 'The Plough' as late as the end of 1917. The first inkling that RAF Shotwick had become operational was a fine period in June 1918, when there were easterly breezes and all day long the air was filled with a humming sound which mystified people completely, until one evening, after staring intently across the Dee in the Shocklach direction, six or seven black specks could be seen moving steadily in random formation across the sky. They had just turned north at Bangor church tower and were making for base on the last leg of a cross-country map-reading exercise.

This was quite civilised, but a ruder introduction to the RAF awaited country folk as fields and ponds began to be shot up by observers/gunners under training. Generally this was confined to the wider open spaces with a greater density of ponds on the Cheshire side of the Dee around Shocklach Hall and St. Edith's Church, the latter a prominent landmark standing alone in its fields. But occasionally ponds on the Denbighshire bank, between Cobham and Sutton Green, also 'copped it'. As observers forsook the comfort of their little 'office' and braced themselves against the wind to load their Lewis gun or to exchange drums without dropping anything overboard, the pilot's gloved hand would vaguely indicate a pond or group of ponds, each 'looking no bigger than a 5s. piece', and down they would go. The series of water-spouts from each burst of the machine-gun may have played havoc with pond life (if any), but they were a sure-fire indication of the gunner's accuracy! This would go on until three or four drums of ammunition had been expended, and did little to endear the RAF to those unfortunates selected as 'target for the day'.

While the observer/gunner and pilot of, say, a two-seater Sopwith

1½-Strutter, could each give their undivided attention to matters in hand, it could not always be so with single-seater fighters. Thus at 07.30 hrs. on Tuesday, 6 August 1918, 2/Lt. J. P. Ruscoe, 20, of Vancouver, British Columbia, attached to 4 TDS, Hooton Park, was shot up by fellow pilot 2/Lt. F. W. Parker. Both were flying Sopwith Pups armed with a synchronised Vickers .303 machine-gun on top of the fuselage forward of the cockpit, firing through the arc of the two-blade propeller. Unfortunately both had selected to shoot up the same pond at the same time, Ruscoe diving from 2,000 ft. and Parker from 800 ft. It was not until he had landed back at Hooton that Parker realised that something was amiss and that he had emptied half of his magazine of 97 bullets into Ruscoe's plane. He admitted to seeing something — he knew not what, such was his concentration — out of the corner of his eye. Ruscoe sustained a bullet wound below his right eye, which, if it had not killed him outright, certainly knocked him unconscious. He received multiple injuries to his skull as his Pup (B7513) crashed, which in any eventuality would have proven fatal.

Other similar incidents could not necessarily be blamed on inexperience or youthful bravado. 'Gremlins', a peculiarly Second World War phenomena, were also busy in the earlier conflict and plagued both those with pips and crowns on their sleeves as well as lowly aircraftmen. Two weeks after the armistice, before the inevitable run-down in the training programme, Capt. P. W. S. Bulman, MC*, AFC, a Western Front veteran being 'rested' as a squadron commander at 51 TDS Shotwick, shot off his own propeller whilst demonstrating to advanced flying pupils the quickest way to empty a military duck pond. He had sufficient skill (and height) to land safely in the nearby field, thus preserving for posterity and the Second World War an experienced test pilot of some renown, first with the Royal Aircraft Establishment, Farnborough (1919-25) and, from 1 July 1921, as Chief Test Pilot with H. G. Hawker Engineering Co. (perhaps better known from 1933 as Hawker Aircraft Ltd.), as well as occasionally testing prototypes for other companies such as Blackburn Aircraft & Motor Co., Ltd. F/Lt. Bulman tested inter alia the Hawker Hornet, Fury, Hart, Hind, Audax and, on 6 November 1935 at Brooklands, K5083, the prototype of the Hawker Hurricane!

Paul Ward Spencer Bulman had gained his Military Cross in 1917 for sustained low level bombing attacks on enemy positions and transport on four successive days. Out of sixteen bombs dropped, he scored eight direct hits and planted seven others with 10 yards of target. He could usefully have taught Bomber Command something! In 1918 he was awarded the Air Force Cross and a bar to the M.C. Following his experimental and testing work he was awarded two bars (1921, 1923) to the A.F.C. Until the George Medal was instituted in September 1940 there was no appropriate gallantry award for test pilots. The A.F.C. was, and still is, awarded for their often hazardous work. Bulman retired from the RAF with the rank of Group Captain and a well-earned C.B.E. (1943), having in 1941–2 headed the Aircraft Test Branch, British Air Commission, Washington. 'Lady Luck' had indeed smiled upon him that rather fraught day over Puddington, Wirral, back in November 1918!

At best, the rough quadrilateral Sealand-Bangor-Marbury-Beeston Castle-Sealand was quite an endurance test, given current standards of navigation. But with unreliable machines many emergency landings were made at Cacca Dutton, The Bryn, Little Bryn in fact any farm in Isycoed parish and across the Dee in Stretton, Caldecott and Shocklach that had fields of about twenty acres or more and looked flat from the air! The tiny township of Grafton, a mere 395 acres, between Tilston and Castletown, seemed to be specially favoured by pupil pilots in distress, a preponderance of medieval ridge and furrow notwithstanding! This was not because historically Grafton was one of Cheshire's mysterious 'extra-parochial' places, but rather because there was one field in particular, bereft of ponds, boggy patches and other hazards, an almost perfect oblong with just a two-foot fall between boundary hedges, that offered a 1400–1650ft take-off run, the latter distance corner to corner. Such landings annoyed farmers no end, not on account of the pilots with whom they had every sympathy in their predicament (shooting-up of duck ponds notwithstanding), but because of neighbours and locals who should have known better — who magically materialised from nowhere, crashing through hedges, breaking down fences and trampling crops in their eagerness to view the offending aeroplane close to.

Nonagenarian Harry Brereton of Lowcross Hill, Tilston, vividly recalls his 'first plane down' on Grafton field: '… in no time at all the pilot had a number of young helping hands ready to give assistance. They came from all over the place! The Tilston police constable made arrangements for the local Army Volunteers to guard the aircraft whilst the pilot was put up at the 'Carden Arms' in the village. As can be imagined in such a quiet backwater, the tales he told after a few drinks spread round the community like wildfire and he became quite a local hero…', which might explain the rather recalcitrant attitude adopted the next day.

Next day there was little work done anywhere as a salvage party — in khaki uniform arrived to inspect the plane and make preparations to fly it off: '… The pilot then arrived from Tilston and complained bitterly about the reliability of his machine. Apparently this was his third or fourth forced landing. This time it was one of the large valve springs on top of the cylinder head that had caused the trouble…'

The pilot was refusing to fly off the field when two 'high-ranking' officers (a captain and a lieutenant) arrived in a Crossley 'Tourer' and '… read the riot act! This seemed to bring the pilot to his senses and he duly taxied the machine to the far corner of the field. The party of airmen held the tail down while he revved up the motor. The next moment the machine sprang forward, climbed into the summer sky, circled the field once, the pilot waving to us and giving the thumbs-up sign to the ground crew, and flew off in the Farndon direction. It was all over in a few moments…'

In the calendar of his mind Harry Brereton is adamant that military flying on the Cheshire side of the Dee commenced as early as 1915 — which ante-dates the establishment of RFC training units at Shotwick, Shawbury, Tern Hill and Hooton Park by some two years. His oral testimony may therefore shed some light on the operation of early civilian flying schools, such as that of Thomas Murthwaite Dutton at Queensferry, within Chester District, Western Command, a necessary provision at that time for those seeking admission into the RFC before the latter's own tutorial infra structure was in place.

Over the next few months aircraft continued to drop in with monotonous regularity, as Harry Brereton recalls: '… winter came and went and then we were back in the summer of 1916 [sic] … One morning the lads came running into the village to tell of an aeroplane standing on its nose in Grafton Field. It was 'The Charge of the Light Brigade' all over again, back down to the fields to see the new arrival. This time it was planted in the ground to just past its engine. It was a bigger aircraft than before, and was painted with very bright roundels. It had two people in it; they had survived the crash with just a few scratches and bruising. Help was on its way in no time as, in order to attract attention, the flyers had fired a Verey pistol which they later showed to us awe struck lads. . . '

The removal of the aircraft took a few days as it had to be dismantled, the engine dug out, and loaded onto a lorry, '… no mean feat for those times when main roads were very narrow and the country lanes no more than tracks… ' Such, eighty years on and making allowances for possible slips in the dates, are one boy's recollections of the seasonal disruptions to the ordered rural round in Tilston parish during the First World War.

Things were equally exciting on the Denbighshire side of the river. One evening in July 1918 a Sopwith Dolphin, possibly from Hooton Park, pancaked onto a Cacca Dutton Farm clover field just as a farm worker was going home. That night in 'The Plough' he was full of his encounter with the aviator and gave his impression of the pilot: 'A very nice young chap he were — Hamerican hor Canhadian hofficer, well educated I should think. The fust thing he sed to me as I came up to 'im was "Were be I?"'

Three days later another Dolphin landed in the very same clover field! Must have been a Thursday, for it emptied tiny Bethel chapel by the smithy down Redwither! Again farmer and son were too late to prevent onlookers further damaging the crop. The pilot made two attempts to take off, the crowd following the machine sheep-like as it taxied from one end of the field to the other. At the second attempt the aircraft reached a height of about 100ft before the engine failed again and the machine came down two fields away, this time in a field of wheat! The pilot checked his engine and shouted for a dozen or so strong youths to hold onto wing struts and tail plane whilst he revved up the engine. The flattening of the wheat was of no import! This was a matter of national importance! Everything seemed to be alright and the Dolphin managed to take off once more. But perhaps the drag of the corn and the great cloud of dust proved too much as, with a lot of spluttering, the engine coughed once or twice and cut out. The aeroplane came down in the next clover-field. Now at least one person's worries were over. This field belonged to The Bryn!

There were many similar occurrences right up to the end of the war. Every time an engine cut out work stopped and heads popped out of doors and windows, wondering where it was this time. Almost all the pilots spoken to were 'townies', who had little idea of the type or importance of the crop into which they had just ploughed. In any case, they had other things to worry about. On one occasion a young lad guided a pilot to The Vicarage, Bowling Bank, to use the telephone and was most concerned to hear him tell the Chief Flying Instructor at Shotwick that he had come down in a clover-field! 'No, no!', the lad interrupted, 'it's a wheat field!' He knew the pilot could never hope to take off from where he was. The corn was ready for harvesting, and as the aeroplane landed only its upper wing was visible as it surged forward with its idling propeller throwing up ears of wheat like some giant catherine wheel!

Some four years later 5 FTS Shotwick maintenance recovery log for 22 September 1922 throws up a similar accident in almost the same spot — a DH9A (E924) was written off after engine failure and a forced landing in a field of standing corn in a cross wind. The aircraft was wrecked and so was much of the crop. Records give the incident at 'Dutton Dithers', which being interpreted was the small detached part of Dutton Diffaeth township on the north bank of the Clywedog river between Pickhill Bridge/Bowling Bank and Redwither which largely disappeared under the Second World War Royal Ordnance Factory. Recovery records are not so precise as to where another 5 FTS aircraft., this time Avro 504K serial H2424, came to grief on 13 June 1924, again in ripening corn, and sustaining serious damage to undercarriage and propeller.

Such emergency landings were made by pupils in early stages of training in generally unreliable machines. They took willy-nilly to the air with that eternal optimism of youth. It is surprising that there were so few fatal accidents. When they did occur it was, more often than not, civilian bystanders who suffered horribly. Thus, on Wednesday, 19 June 1918, at Holdgate, near Much Wenlock, one Harriet Bounds, 18, was virtually decapitated when struck by the propeller of a plane which had made an emergency landing in Lowe's Field. Flight Cadet Arthur Dunn had taken off from 'Shrewsbury Aerodrome' (*i.e.* Shawbury) and at about 10.10 hrs. ran into engine trouble which forced him to land. The field was not level and the machine ran the whole length crashing through some bushes before coming to a halt. Without the pilot realising it, his undercarriage had twisted.

After tinkering with the carburettor Cadet Dunn was ready to take off. There was no one to hold back the crowd that had gathered, but he had at least managed to clear a path to his front. But as soon as it started the machine twisted round on its left wheel and tilted on to its nose pinning the young girl underneath. The plane could not be shifted and the emergency services (doctor and local police sergeant) were sent for. Inexperienced flyer the pilot may have been, but he had three previous forced landings under his belt! Although the aeroplane had an 80 h.p. engine, no precautions were taken to hold down the tail 'as he was only on half throttle' and was fully expecting to surge ahead across the field, but …!

Of all those that landed in the 'Bermuda Triangle' between Holt, Erlas and Bangor, none reported the slightest injury. This was all the more remarkable if one considers the hazards of landing in a hedged field with ditches and tall crops hiding possible obstructions not readily visible from the air. The early biplanes were small, resilient and light and became airborne without much difficulty. Only on one occasion does folk memory recall a pilot balking at the prospect of flying his Avro 504 from a small ploughed field at Cae Mynach, and this despite an harangue and threats of a court martial from a very superior officer! In the event it was now up to the instructor. To maximise his run he bounced diagonally across the furrows and, after a few anxious moments cleared the hedge with inches to spare, to loud cheers and much cap waving (and perhaps one or two sighs of disappointment?) from the large crowd that the bush telegraph had summoned to witness the take-off. Many had cycled from Wrexham and Holt.

These were great days for country lads interested in aeroplanes, especially one who caught the aviation bug after watching one Captain Brereton struggle in gusty conditions to master his Blackburn Mercury monoplane at the Wem Agricultural Show in 1913, and who, thirty years later, would gain his blue, red and gold 'Spitfire' proficiency badges (aircraft identification) with the Royal Observer Corps at Aldford. In 1918 planes were all shapes and sizes and did not fly so high or as fast. The roll-call was endless, from vintage Maurice Farmans sailing slowly over the tree tops, FE2b 'pushers', BE2cs, Pups, Dolphins, Camels, S.E.5s, 504Js and Ks to the daddy of them all, the giant Handley Page 0/400 bomber which latter could only have come from Tern Hill.

In 1940 the many fields on which these First World War 'stringbags' came to grief disappeared for ever beneath the buildings, bunkers, railways and marshalling yards of No. 35 Royal Ordnance Factory, now in turn swallowed up by the Wrexham Industrial Estate. But even during the Second World War the place would appear to have exerted the same fatal attraction for a new generation of pilots and aircraft!

So it was that, at 0915 hrs. on 5 October 1943, a 17 (P)AFU Miles Master II, AZ593, from RAF Wrexham, piloted by 18-year old Sgt. C. F. Lee RCAF, came to grief in a field then belonging to Marshley Farm, Ridley Wood but now lost to the Owens Corning Fibreglass factory. He was 55 minutes into his solo training flight involving local map reading and some aerobatics. At Bangor Bridge and church he had just turned onto a northerly course. In practising some steep turns and spins he apparently came too slowly into a turn and stalled. The Master spiralled uncontrollably into the ground just outside the ROF boundary fence, killing its pilot instantly.

Half a century later John Crosby recalls the incident vividly, being taken as a child to the scene of the crash by his father, a ROF security man: 'Beyond a large haystack lay the Master, surrounded by a small group of people. Its nose had crumpled right back to the cockpit. It had ploughed a broad furrow across the field and pushed up a large pile of earth in front of it. Its wings were still on, albeit damaged a lot'. The eight-year old stood wide-eyed, open-mouthed, listening to the muttered comments of the bystanders. The pilot had been taken straight to the mortuary at Wrexham Cemetery, where he would be buried with full military honours two days later.

The reminiscences continue: 'I walked all round this war plane. It was a sight I'll never forget. The men invited Dad to put me inside the glass cockpit if I wanted to go. I did not need a further invitation. I was on the wing and climbing into the seat in the front. All the dials and clock glasses were broken and the front windscreen was missing in one corner. When I look back now and remember the mess inside I think "Just 3/4-hour i earlier a young airman met his death in the same seat… " But as a child you never gave it another thought… '

An even more bizarre incident involved Master II AZ698, again from 17 (P)AFU, Wrexham, which dived vertically into the only surviving patch of green sward on the ROF site, to the rear of what was later British Tissues Ltd., but which was then the cordite baking, rolling and cutting plant! The ROF lay smack in the middle of RAF Wrexham's 'high activity, low flying' area. AZ698, pilot F/Sgt. R. Kershaw, pupil Sgt. H.W. Scott RCAF, had taken off from Borras at 1315hrs on 16 September 1943 for a 45–minute training flight with elementary aerobatics on circuit. The Master gained height normally, entered a patch of low cloud and promptly collided with Wellington X3807 from 23 OTU at Pershore, Worcestershire, which had unknowingly drifted too close to Wrexham's training circuit. Its navigator had not obtained a proper route forecast and one of the recommendations from the subsequent Court of Inquiry was that NAVEX briefing sessions should stress and pin-point the location of airfields lying on or near proposed routes. But this would come too late for the unfortunate crew of AZ698. The Wellington trundled on, its crew quite unaware of the collision! X3807 had been delivered from Vickers, Squires Gate, to 150 Squadron at Snaith, Yorkshire, in the Spring of 1942, taking part in night bombing raids. It was transferred to 23 OTU when 150 Squadron left for Algeria at the end of the year. It had a long service career, later with 83 OTU at Peplow, before finally being struck off charge on 12 May 1947. The Gods were very kind over Holt that Friday in 1943.

Out in Common Wood, Derek Parker, Plas Devon, and his mates watched AZ698 take off from RAF Wrexham as they had watched hundreds of others before. The day was fine, but with some broken cloud. As the tiny aircraft climbed eastwards their attention was suddenly taken by a larger aircraft approaching from the south-east. The unfolding drama is freshly recalled: 'At first we did not realise the dangerous situation developing in front of our eyes. Up and up the little aircraft went, and before you knew it, we were all shouting 'They are going to hit each other!' We saw the larger aircraft go into a small cloud-bank. A second later the small aircraft also entered the same cloud. We could see them just for a moment as the smaller one disappeared under the big one. It re-appeared and hung there for a split second. Then we could see bits fall from it. The plane went into a spiral, gaining speed as it fell to earth. Black smoke began to pour from it. It was a sickening sight. Your stomach felt like lead watching helplessly as we did. No parachutes appeared so you knew that some poor lad was going to his death. Big parts fell away, then it went out of sight behind some trees on the horizon, somewhere in the direction of Bryn Farm or the ROF … We then looked for the big one. We could hear it just above us … and it just kept on flying away. It sounded OK and gone in a minute … It made me feel quite sick for days after, and the memory has not faded. It's just like yesterday. I remember the funeral of the one lad … We got used to seeing low flying all over Wrexham and district, with aircraft getting into trouble and landing everywhere, but this one was different!' Sgt. Howard William Scott RCAF, 25, lies buried in Wrexham Cemetery, far from his native Toronto.

SOME ACCIDENTS AND CASUALTIES, 1916–19

22.10.16 Llangefni. 'Submarine Scout' type blimp SS18. Blown out to sea whilst trying to land. Heavy landing in water; envelope and car broke in two. Engineer drowned. Pilot, Sub-Lt. A. Thompson rescued by passing ship.

2.1.17 Luce Bay. 'Submarine Scout' type blimp SS23. Caught in gale, landed on beach, Red Wharf Bay, deflated after 3 hours.

-.4.17 Llangefni. 'Submarine Scout' type blimp SS33. E/F, forced landing Cemlyn Bay, north coast of Anglesey; slight damage.

9.6.17 Tern Hill. Airco DH2 6008. 2/Lt. Edward Phillip Hughes, RFC, 24, 10 TS. Ellesmere, Cape Province. Died 27. 6.17 from injuries received 'falling with a flying machine'. Military funeral at Tilstock.

16.7.17 Tern Hill. Airco DH2 A2560 2/Lt. George Everest Cayford, 21, Wanstead. Artists Rifles. Killed. Solo flight. 09.30 hrs. took off; turned without banking, side slipped into ground.

20.9.17 Tern Hill. Cadet E. J. C. Treadwell, 22, AFC, St. Kilda, Victoria, 30 TS Aust. Making instrument check, jumped off wing into propeller. Buried Tilstock cemetery.

2.10.17 Shawbury. Sopwith Camel. 2/Lt. A. C. Tallent, 10 TS. Off course, engine trouble, forced landing Overton (Malpas) 'wrecking the propeller and damaging the sails '.

10.10.17 Tern Hill. Sopwith Camel B6222. 2/Lt. C. E. Rider, RFC. Killed. Spun into ground from 3,000 ft.

13.10.17 Shawbury. Sopwith Camel B5164. 2/Lt. Harold James Cryer, 19, Herne Hill, London, 10 TS. Killed. Biplane observed at 1,000 ft., spun in nose dive flattened out at 500 ft., lost altitude, into nose dive from which he did not recover.

15.10.17 Tern Hill. Airco DH5 9234. 2/Lt. Herbert Frederick Mayer (Meyer), 29 Canadian. Killed 09.30 hrs. flying over Stoke upon Tern spinning nose dive into ground, one wing falling off. Engine buried 2ft in ground. Buried in Stoke upon Tern cemetery.

23.10.17 Tern Hill. Nieuport 20 A6741. 2/Lt. Leslie Thomas Hopkin (Hogben), RFC, 18, killed outright in mid-air collision with 2/Lt. F. Jickling at 11.00 hrs. near Stoke on Trent.

23.10.17 Tern Hill. Avro 504A B3105. 2/Lt. Frank Jickling, RFC, 22. Injured in mid-air collision with above at 11.00 hrs. near Stoke on Trent. Died later same night; buried in Stoke upon Tern cemetery.

23.10.17 Tern Hill. Airco DH5 A9467. 2/Lt. Sydney Harold Smith, RFC, 24. Engine failure, stalled at 150 ft., spun into ground, fatal injuries. Buried in Stoke upon Tern cemetery.

4.11.17 Shawbury. Sopwith Camel B6252. 2/Lt. Albert Cecil Tallent, 10 TS. Killed. Pilot choked engine, during turn into aerodrome lost speed, went into spinning nose dive. 'No reflection on instruction received'.

7.11.17 Llangefni. Airco DH4 A7654. 2/Lt. Bernard Carter. Killed. Ferry flight (from 2 AAP, Hendon). Gust of wind or stalled on turn, crashed on landing. Observer, Cpl. Harold Smith seriously injured.

7.11.17 Llangefni. Airco DH4. Forced landing in bad weather on Lavan Sands (low tide). Plane destroyed by tides, engine salvaged.

14.11.17 Tern Hill. Maurice Farman S.11 Shorthorn B4787 Lt. R. Wynford Phillips, killed, another officer 2/Lt. John William Barr, injured, when plane crashed to ground and caught fire.

23.11.17 Llangefni. 'Submarine Scout' type blimp SS25. Forced landing, car and elevators damaged.

10.12.17 Tern Hill. Unidentified aircraft. Pilot, from 34 TS, killed. Probably fainted at 3,000 ft., lost control, spun in.

13.12.17 Tern Hill. Sopwith Camel B6445. 2/Lt. D. G. Scott, RFC, 21. Canadian. Killed. Buried in Stoke upon Tern cemetery.

18.12.17 Tern Hill. Sopwith Camel B6268. 2/Lt. Robert Ernest Cleary, RFC, 22, Canadian. 30 TS. Killed. Engine failure, flat turn, spun in. Buried in Stoke upon Tern cemetery.

21.12.17 Shawbury. Sopwith Camel 9B B2310. 2/Lt. J. F. Kneale, RFC, 10 TS Shawbury. Lost in fog, crashed. Buried Tilstock.

3.1.18 Tern Hill. Sopwith 1¹/2 Strutter A5964. 2/Lt. Andrew John Cumberland (Cumbernald), RFC, 20, Canadian. 43 TS. Killed Structural fault, pieces fell off plane, tail 'hung down', crashed at Tiverton, near Tarporley. Burst into flames pilot found under engine, burnt in head and legs. Buried in Stoke upon Tern cemetery.

5.1.18 Shawbury. Sopwith Camel B6305. 2/Lt. J. S. Woods. Injured. Pilot 'error of judgement'.

20.1.18 Tern Hill. Sopwith Pup B6089. 2/Lt. Reginald James Thomas Forsyth, 25, AFC, from Sydney, NSW. 6 TDS. Air sickness whilst rectifying position in formation, banked too steeply, spun in from 1,000 ft. Fatal injuries to head and ankles. Died Prees Military Hospital 16. 2. 18. Buried Tilstock.

27.1.18 Tern Hill. Sopwith Camel B6433, belonging to 6 TS AFC. Forced landing, tipped on to nose.

27.1.18 Shotwick. Sopwith Pup B5792. 2/Lt. John Daniel Brendel, RFC, 22, Canadian. Killed. Flying 'Scout' machine over Mersey at 6,000 ft. Engine failure ('reports like a revolver'), nose dived into Mersey off Eastham. Body trapped inside, recovered at low tide. Buried in Shotwick churchyard.

5.2.18 Tern Hill. Sopwith Camel B6442. 1st Air Mech. H. T. A. Garner. Injured. Propeller accident.

10.2.18 Shotwick. Avro 504 D12. 2/Lt. A. L. Fachnie, 90 Sqn. Minor injuries. Engine failure, returned to airfield, misjudged distance, crash-landed.

10.2.18 Tern Hill. 1130 Corp. A. Morgan, AFC, 31, 6 TS. Killed. Propeller accident. Buried in Tilstock cemetery.

16.2.18 Tern Hill. 2/Lt. R. J. T. Forsyth, AFC, 6 TS. Killed. Stalled on turn, spun in. Buried in Tilstock cemetery.

18.2.18 Tern Hill. Bristol Scout D B716. 2/Lt. W. J. McGinn, RFC, 20. Canadian. 34 TS. Killed. Buried in Stoke upon Tern cemetery.

21.2.18 Tern Hill. Sopwith Camel B6442. 2/Lt. L. W. Heseltine. Injured. Stalled on take off, side-slipped in to ground.

25.2.18 Shawbury. Sopwith Camel B4643. 2/Lt. E. R. Watt, 67 TS. Killed. Stalled, spun in.

28.2.18 Tern Hill. Sopwith Camel E7312. 2/Lt. G. F. Hubbard, 43 TS. Injured. Stalled on turn near the ground, spun in.

-.3.18 Shawbury. Sopwith Camel B6416 ('The Straffer') belonging to 10 TS. Fuselage, wings and tail-plane covered in War Bonds promotional advertising. Engine failure, forced landing in Ludlow Park.

9.3.18 Tern Hill. Sopwith Camel B6434. 2/Lt. P. W. Atkins, 34 TS. Injured. Lost control, spun in, colliding with Camel B4607 and an Avro 504J.

16.3.18 Llangefni. 'Sea Scout Pusher' type blimp SSP6. Engine failure. Abandoned at sea, landed at Chichester.

21.3.18 Shawbury. Sopwith Camel B2521. Lt. Charles McElroy Carpenter, RFC, 21, Canadian (b. Hamilton 17.11.1896), 67 TS. Killed. Buried in Shawbury churchyard.

-.4.18 Shawbury. Sopwith Camel B9316, of 10 TS, crashed Nuneaton.

1.4.18 Malahide. 'Submarine Scout Zero' type blimp SSZ50. Night landing when deflated.

2.4.18 Llangefni. 'Submarine Scout Zero' type blimp SSZ51. Envelope torn by propeller, deflated.

8.4.18 Shotwick. Sopwith Dolphin C4140. 2/Lt. B. S. Crecine, 90 Sqn. Minor injuries. Engine failure in take off, stalled, aircraft badly damaged.

22.4.18 Shawbury. Airco DH6 C7215. Capt. Samuel Trahern Bassett Saunderson, RAF. Killed. Buried in Shawbury churchyard. Capt. Norman Victor Harrison also killed.

25.4.18 Shotwick. Avro 504K C599. 2/Lt. John Jewett Miller, 25, Mass. (USA). 95 Sqn. Killed, buried in Shotwick churchyard.

26.4.18 Llangefni. 'Submarine Scout Zero' type blimp SSZ35. Engine failure, drifted, valved gas, taken in tow by trawler, dumped on beach at Llandudno to await repair party.

27. 4. 18. Shotwick. Sopwith Pup B1742. Lt. Victor William Vallette Lowrie, RAF, 18, Radyr, Cardiff, flying instructor. Hit air pocket, spun in at low altitude. Died later of shock and injuries at Chester 'War Hospital'.

5.5.18 Shawbury. Airco DH9 C6657. 1st Air Mech. C. R. Clack, 131 Sqn. Thrown out and killed in forced landing. Pilot injured. Buried Tilstock cemetery.

18.5.18 Shotwick. Sopwith Camel C112. Sgt. C. V. Simmond, 67 TS. Seriously injured. Engine failure at 50 ft., stalled, spun in.

22.5.18 Hooton Park. Sopwith Dolphin C3915. 2/Lt. Kenneth Alonzo Nelson, RAF, 22, Howard City, Michigan. Spun in, died of injuries 23. 5. 18. Buried Chester General Cemetery.

22.5.18 Shotwick. Avro 504K D6292 of 90 Sqn. Engine failure, forced landing, tipped on nose in soft sand on edge of Dee estuary.

25.5.18 Shawbury. Airco DH9 D5613 2/Lt. George Roper, RFC, 25, Heubenville, Ohio. 9 TDS. Killed flying at Sealand; buried in Shawbury churchyard.

27.5.18 Hooton Park. Sopwith Pup D1429. 2/Lt. Walter Scott Murray, RAF, 20, Gloucester, New Jersey. 4 TDS. Killed in flying accident. Buried in St. Mary's churchyard, Eastham.

1.6.18 Shotwick. Sopwith Camel C77. 2/Lt. F. Shepherd, 67 TS. Seriously injured. Bounced on landing, stalled, crashed on to starboard wing.

2.6.18 Hooton Park. Sopwith Pup B6148. 2/Lt. Clarence Scott Garden, RAF, 26. Seaforth, Ontario. 4 TDS. Killed, buried in Eastham churchyard.

5.6.18 Shotwick. Sopwith Camel C8304. 2/Lt. E. S. Hart, 96 Sqn. Killed. Aerobatics, into a spin at 2,000 ft., nose dived into ground.

12.6.18 Shotwick. Sopwith Pup C278. Lt. Harry Nelson Hastie, RAF, 24. Ontario, Canada, 95 Sqn. Killed, buried in Shotwick churchyard.

16.6.18 Shawbury. RE. 8 D4985. 2/Lt. A. G. McGillivray, RAF. 131 Sqn. Killed in flying accident. Buried in All Saints' churchyard, Siddington, Cheshire.

16.6.18 Shotwick. Avro 504K B8780. Lt. Hugh Robert Fuhr, RAF, 19. Cape Province. 67 TS (ex-12th South African Infantry). Killed. Buried in Shotwick churchyard, later exhumed.

18.6.18 Shotwick. Sopwith Camel C8305. 2/Lt. G. C. R. Hamilton, 96 Sqn. Killed, stalled on turn, spun in from 2,000 ft.

19.6.18 Shawbury. Flt. Cdt. Arthur Dunn. Forced landing at Lowe's Field, Holdgate, Much Wenlock. Swung and tipped on nose on take off, killing 18 year-old girl.

20.6.18 Hooton Park. Avro 504J B8064. Capt. William Reginald Guy Pearson, 21, London; Lt. William MacFarlane M.C. (5 Batt. Royal Scots), 23, Edinburgh. Both killed in mid-air collision with Lt. Flynn. Pearson buried in St. Mary's churchyard, Eastham; MacFarlane in Dean Cemetery, Edinburgh.

20.6.18 Hooton Park. Avro 504J D7562. Lt. Vincent Flynn, 20, American, New Jersey. Killed in mid-air collision with above.

25.6.18 Hooton Park. Sopwith Dolphin C3997. 2/Lt. Oscar Oliver Mousley, RAF, 27. Weston, Ontario. Killed, buried in Eastham churchyard.

28.6.18 Shotwick. Sopwith Dolphin C3876. 2/Lt. B. Green. Minor injuries. Bad landing, left wing down, left wheel came off, tipped on nose, aircraft badly damaged.

30.6.18 Shotwick. 2/Lt. G. W. Sweeney, 67 TS. Seriously injured. Aircraft stalled at 100 ft., spun in.

9.7.18. Shotwick. Sopwith Pup B5399. 2/Lt. Horace Edgar Kingsmill Bray, RAF, 22. London, Ontario. 67 TS. (7th Canadian Mounted Rifles, CEF). Killed, buried in Shotwick churchyard.

12.7.18 Shotwick. Sopwith Camel D8127. Flt. Cdt. T. A. Emmerton, 55 TS. Injured. Aerobatics, into spin in cloud base. Crash landed.

16.7.18 Hooton Park. Sopwith Pup D4159. 2/Lt. Laurence Phelps Waite, RAF, 23. 4 TDS. Ingersoll, Ontario. Killed, buried in Eastham churchyard.

21.7.18 Shotwick. Sopwith Camel B9224. 2/Lt. C. W. Duncan, 67 TS. Injured. Engine failure, dived into ground

21.7.18 Hooton Park. Sopwith Dolphin C4138. 2/Lt. J. P. F. English, RAF, 25, Troy, New York. Killed in flying accident. Buried in St. Mary of Angels R. C. churchyard, Hooton. Body later exhumed.

28.7.18 Shawbury. BE 2e A1391. 100546 Flt. Cdt. Clarence Edward Brown, RFC, 18. 9 TDS. Killed. At 3,000ft got into a spin, crashed, burst into flames. Verdict: 'Lost his head and failed to shut off engine. ' Buried in Shawbury churchyard.

29.7.18 Shawbury. RE. 8 C2397. 2/Lt. Joseph Arthur Freeman. Killed. 35 flying hours. Normal take off, spun wing over wing, crashed, petrol tank blew up, pilot badly burnt. Died later of injuries.

30.7.18 Tern Hill. Avro 504K D6361. Major Cuthbert Everard Brisley, 32, CO 13 TDS, Tern Hill; 155338 (Boy) Pte. 2nd Class Fred Lythgoe, 18, Atherton, Manchester. Both killed. Shortly before 12.00 hrs. plane at 6,000 ft., pilot fell out, passenger killed in crash. Brisley is buried in Market Drayton cemetery.

31.7.18 Shotwick. Sopwith Camel E9973. Flt. Cdt. G. McLeish, 51 TDS. Killed. Pilot fainted at height, lost control, dived into ground at Saughall.

5.8.18 Shotwick. Sopwith Dolphin D3698. 2/Lt. Allan John Brian Meikle, 19, Andoversford, Gloucs. Killed. At 10.40 hrs. took off too steeply, flat turn to port, stalled at 250 ft., burst into flames.

6.8.18 Hooton Park. Sopwith Pup B7513. 2/Lt. James Frank Roscoe, RAF, 19. Canadian, Vancouver (formerly of Liverpool). 4 TDS. Killed. At 07.20 hrs. got into line of fire of another plane during target practice, bullet wound below right eye, fatal injuries on crash. Buried in Eastham churchyard.

11.8.18 Shotwick. Sopwith Camel F1516. Capt. H. Brokenshaw, 51 TDS. Seriously injured. Spun into ground during firing practice.

11.8.18 Shotwick. Avro 504K E1719. Lt. Leonard S. Morange, RAF, 22. Bronxville, N. Y. 51 TDS. Killed in mid-air collision. Buried in Shotwick churchyard. F/Cdt Robert Oughtibridge, 18, Leeds, also killed; buried in Lawnswood Cemetery, Adel, Yorks.

11.8.18 Hooton Park. Avro 504K E2932. Lt. Malcolm Saunders Pettitt, 19, Orpington, Kent; 4 TDS. Flt. Cdt. C. B. Cooke in back seat. According to local press Both killed in mid-air collision with Flt. Cdt. Lund (below). Tail struck, turned upside down, glided to earth, hit tree in wood, telescoped. According to F. S. Form 559 Cooke was injured. Pettitt died from shock and multiple injuries on way to hospital.

11.8.18 Hooton Park. Sopwith Pup D4106. Flt. Cdt. William Peedie Lund. 4 TDS. Solo practice flight. Seriously injured (?killed). Damaged wing in collision with above, spun to earth in field 300yds away from above crash.

13.8.18 Shotwick. Sopwith Camel D8129. Lt. J. J. A. Ince, 55 TS. Injured. Engine failure, stalled on approach to aerodrome.

14.8.18 Llangefni. Airco DH6 C2021. 521 Flt/255 Sqn. Failed to return from sea patrol.

14.8.18 Bangor. Airco DH6 B3021. Lt. J. R. Johnstone, 255 Sqn. Killed. Engine failure coming in to land after patrol. Crashed.

22.8.18 Shotwick. Sopwith Camel E1462. F/Cdt. C. W. Joughin, 51 TDS, slightly injured. Stalled on turn and spun in.

23.8.18 Shotwick. Sopwith Camel 130/140 h.p. Clerget E1461. 2/Lt. F. R. Orris, 51 TDS. Injured. Stalled on turn after take-off. Crashed.

23.8.18 Bangor. Airco DH6 F3350, 'B' Flt/244 Sqn. Crashed in sea whilst on patrol. Lt. C. W. S. Hall picked up by RN destroyer.

25.8.18 Shawbury. Airco DH9 D5587. 90265 F/Cdt. Arthur Dunn, 18, Manchester. 9 TDS. Engine vibration, forced landing at Ash. Repaired, took off, at 200ft too tight a bank to port, side-slipped in and disintegrated at Prees Heath. Pilot died of injuries next day. Batman 1186648 Pte. Frank Baskerville, 26, Manchester also killed and buried Manchester Southern Cemetery.

26.8.18 Bangor. Airco DH6 B3023. 522 Flt/244 Sqn. Crashed on rigging test. Lt. A. C. Bencher, Air Mech. Nicholls, slightly injured.

28.8.18 Shotwick. Sopwith Camel E1462. Flt. Cdt. C. W. Joughin, 51 TDS. Injured. Stalled on turn, spun in.

30.8.18 Tern Hill. 2/Lt. Howard Knight Stevens, 25, Canadian. 13 TDS. Killed in flying accident. Buried in Market Drayton cemetery.

5.9.18 Hooton Park. Sopwith Dolphin C3998. 2/Lt. Claude Elsdon Elliott, RAF, 22. St. Thomas Ontario. 4 TDS. (ex-Canadian Army Medical Corps). Killed, buried in Eastham churchyard. (C. E. Elsdon on gravestone).

7.9.18 Bangor. DH6 B2971. 521 Flt/244 Sqn. Engine failure, forced landing. Towed back to airfield.

15.9.18 Llangefni. 'Submarine Scout Zero' type blimp SSZ51. Lost in Irish Sea, crew saved by USS *Downs*.

18.9.18 Bangor Airco DH6 B3023. Capt. D. A. Tuck, seriously injured; Air Mech. E. W. Shaw died later from injuries. Stalled on turn, spun in from 60–70 ft.

1.10.18 Shawbury. Avro 504 B8979 (rebuilt). 962953 Pte. L. G. Houseal, American, 9TDS. (F. S. Form 559 as D sect. 262 Aero Sqdn). Killed. 'Sucking in engine', pilot accidentally knocked on contact switch. Deceased struck on head by propeller. Buried in Shawbury churchyard (later exhumed).

2.10.18 Shotwick. Sopwith Camel E7297. 2/Lt. A. L. Fachnie, 51 TDS (second crash). Injured. Spun in on final approach.

4.10.18 Llangefni. 'Submarine Scout Zero' type blimp SSZ50. Damaged by gale.

11.10.18 Bangor. Airco DH6 C2047, 522 Flt/244 Sqn. Lost, crashed into Caernarfon Bay.

12.10.18 Tern Hill. Lt. W. Burt Bickell, 25, Toronto. 13 TDS. Into spin at 200 ft., flattened out but crashed owing to insufficient height. Buried in Tilstock cemetery.

13.10.18 Shotwick. Sopwith Camel E7271. F/Cdt. D. Meyers, 51 TDS. Slightly injuries as hit tree while flying low.

17.10.18 Llangefni. 'Submarine Scout Zero' type blimp SSZ35. Torn envelope, lost in Irish Sea 15 miles NW Holyhead. Crew saved.

19.10.18 Shotwick. Avro 504K H2250. Lt. Herbert Westgarth Soulby, Croix de Guerre, RAF, 22. 51 TDS Killed, buried in Shotwick churchyard. Cdt. H. Beever escaped uninjured.

29.10.18 Shotwick. Sopwith Camel E9982 (built from spares by No. 3 ARD). 2/Lt. Frank Albert Samuelson, 22, Arlington Heights, Mass. 51 TDS. Killed. Pilot 'error of judgement'. Buried in Shotwick churchyard.

1.11.18 Tern Hill. F/Cdt. George Lanncaster Robinson, RAF, 21, PPCLI (East Ontario Regtiment). 13 TDS. Died of pneumonia and complications from earlier accident. Buried in Tilstock cemetery.

9.11.18 Shotwick. Avro 504K F8719. 2/Lt. Francis Athol Hinton, farmer, Auckland, N.Z. 51 TDS. Killed. Engine trouble, lack of flying speed, stalled, and crashed into field at Buerton. Just completed course at Shotwick, about to embark on flying instructor's course. Buried in Shotwick churchyard.

9.11.18 Shotwick. Sopwith Camel F1946. Flt. Cdt. V. E. Gibson, RAF. 51 TDS. Killed, stalled on turn and dived in. Buried in Shotwick churchyard.

10.11.18 Tern Hill. 2/Lt. G. P. Gilliers, RAF. Killed. Engine trouble, dived in. Buried in Tilstock cemetery.

13.11.18 Shawbury. Pte. O. Kious, American. 9 TDS. Killed, buried in Shawbury churchyard (later exhumed).

18.11.18 Hooton Park. Sopwith Dolphin C4011. Capt. Henry Thornbury Fox Russell MC, 21, Holyhead. 4 TDS (ex-RWF). Killed. Experienced flier, flight commander. Presumed fainted or dizzy. Went into spin at 900 ft., did not recover control.

18.11.18 Llangefni. 'Submarine Scout Zero' type blimp SSZ50. Forced landing at Aberaeron, Cardiganshire. Slight damage.

29.11.18 Hooton Park. Sopwith Dolphin C4143. Flt. Cdt. Archie Percival Millhouse, 23, Stretford, Manchester. 3 Sqn. Hooton. Pupil under instruction. Killed. Plane dived from 2,500ft into ground at Church Farm, Capenhurst.

5.12.18 Hooton Park. Sopwith Dolphin C4233. Flt. Cdt. James Wyndham Peddie, 22, Winnipeg. 4 TDS. Pupil under instruction. Killed. Took off, climbed too steeply, stalled at 150 ft., crashed near aerodrome.

14.12.18 Shotwick. Sopwith Camel E1464. Flt. Cdt. Alan Edward Lloyd, 20, Newton Abbot, 51 TDS. Killed. At 10.30 hrs. went into right hand spin from 300 ft., corrected, into left hand spin, crashed on back.

14.12.18 Shotwick. Sopwith Camel F4182 (built up from salvage by No. 3 (Western) ARD). Flt. Cdt. Edward James Rice, 19, Heavitree, Exeter, 51 TDS. Killed. At 11.00 hrs. side slipped, into nose dive, crashed. Pilot fatally injured, died one hour later.

4.2.19 Shotwick. Sopwith Camel D8130. Capt. J. H. Tudhope, 51 TDS. Slightly Injured. Propeller injury.

25.2.19 Monkmoor. Airco DH9A F1179. Lt. Charles Evered Preece, 22, OSRAP, Shrewsbury. Hanley Castle, Upton on Severn. Air test, engine failure, dived in, disintegrated, caught fire. Pilot and 1st Air Mech. (ACl) Harry Welch, 20, Ladywood, killed.

28.2.19 Shotwick. RAF SE5a D6990. 51 TDS. Lost at sea

18.4.19 Short 'Shirl' torpedo bomber N112. Les Parker, Short's test pilot. Uninjured. Crashed into stone wall attempting landing at Holyhead whilst escorting special 'Transatlantic' Shirl, 'Shamrock' which ditched same day.

18.4.19 Short 'Shirl' torpedo bomber Nlll, 'Shamrock' Modified (larger wings and tailplane, immense petrol tank slung under fuselage) for proposed E–W Transatlantic attempt. Crashed in sea off Holyhead en route to Dublin and The Curragh starting point. Major Wood and Capt. Wyllie, RAF uninjured. Aircraft salvaged.

Retrenchment years

Suddenly, after almost two years of near mayhem, aeroplanes ceased to drop out of the sky — or at least not in such great numbers! The years post-1919 would turn out to be years of savage retrenchment and political naiveté, a period when 'cheap' and 'cheapness' were the magic 'in'-words, a decade or so of miserly folly that inflicted irrevocable damage upon all three Services. By 1923 the RAF numbered but 30,188 officers and men compared with the 240,256 personnel on strength in 1918. On paper it could muster only thirty-six operational squadrons, twenty-one of them overseas. With the gunfire of the First World War still reverberating along the corridors of power, these sterile years must be viewed against a common revulsion from the Great War, an implicit trust in universal disarmament, and international security guaranteed by the League of Nations.

Then in 1933 Liberal Britain had reluctantly to acknowledge the threat posed by a Germany that had dropped out of the Disarmament Conference and was beginning to re-arm. Fortunately, despite the weakening effect of the 'Geddes Axe', the paralysis induced by the 'Ten Year Rules', and a mystical adherence to obsolescent theories on prosecuting an air war, there had been almost surreptitious, if not downright covert, moves forward by a small group of dedicated men such as Sydney Camm and Reginald Mitchell. Within the short space of five years their genius, coupled with the advent of the Merlin engine and the American Browning machine-gun, would transform the RAF from a fighting force largely furnished with the low-performance wooden biplanes of 1934 into one equipped with the high-performance all-metal monoplanes of 1939-40. The transition would be given further impetus by the belated application of science to defence, notably Radio Direction Finding (RDF) or radar, and the restructuring of the RAF itself in May-July 1936 into Training, Bomber, Coastal and Fighter Commands. Such are the essential threads to our later narrative, even on a local scale in the so-called 'back areas' of the north-west.

In the early evening of Sunday, 9 March 1919 a disastrous fire struck RAF Tern Hill. Despite the Market Drayton Fire Brigade being on the scene inside thirty minutes and the Shrewsbury Brigade within the hour, a large hangar containing four Handley Page 0/400 bombers and several Avro 504s was destroyed, along with sundry wooden hutments nearby and a timber and canvas Bessonneau hangar containing a quantity of stores. Water had to be pumped from ponds up to half a mile away. The brigades managed to contain the fires, but it was 5 *a.m.* the following morning before the smoking ruins were dampened down. The cause of the blaze was attributed to a small combustion stove in an office attached to one of the hangars.

There was no loss of life, but in the short term the station's usefulness was diminished. This was sufficient to turn the scales in favour of RAF Shotwick when it came to making a decision as to which of the training depôts in the north-west would be retained by the peace-time RAF. The latter site also had the material advantage of a new, as yet unnumbered, Aircraft Acceptance Park attached, just the other side of the railway at 'South' Shotwick, coming on stream only in the closing months of the war. In 1920 Tern Hill aerodrome was sold, but in a sense its training role would continue — but with race horses! For the next fifteen years surviving hangars were used as stables and the airfield circumference as a trotting track or gallop. Other ancillary and former barrack buildings either collapsed through neglect or were demolished. Civilian flying in any form was militated against by the thousands of rabbits whose burrows within two or three years came to undermine the airfield. The full extent of their depredations was not fully realised until contractors moved on site in September 1935. The airfield not only was going to be enlarged, but the whole had to be deep-ploughed, regraded and resurfaced before grassing down. Inevitably subsoil and texture were affected and this

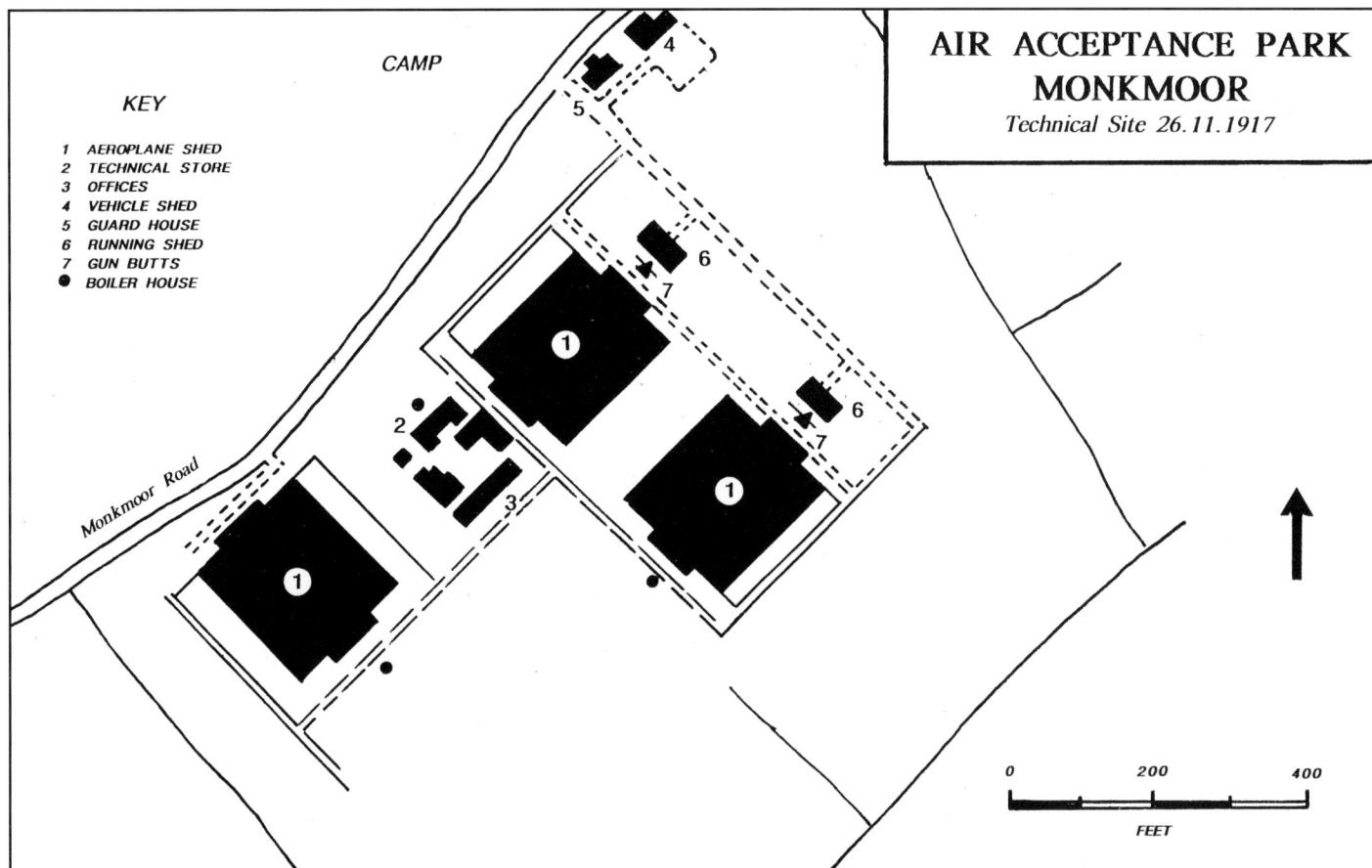

AIR ACCEPTANCE PARK
MONKMOOR
Technical Site 26.11.1917

KEY

1 AEROPLANE SHED
2 TECHNICAL STORE
3 OFFICES
4 VEHICLE SHED
5 GUARD HOUSE
6 RUNNING SHED
7 GUN BUTTS
● BOILER HOUSE

CAMP

Monkmoor Road

0   200   400

FEET

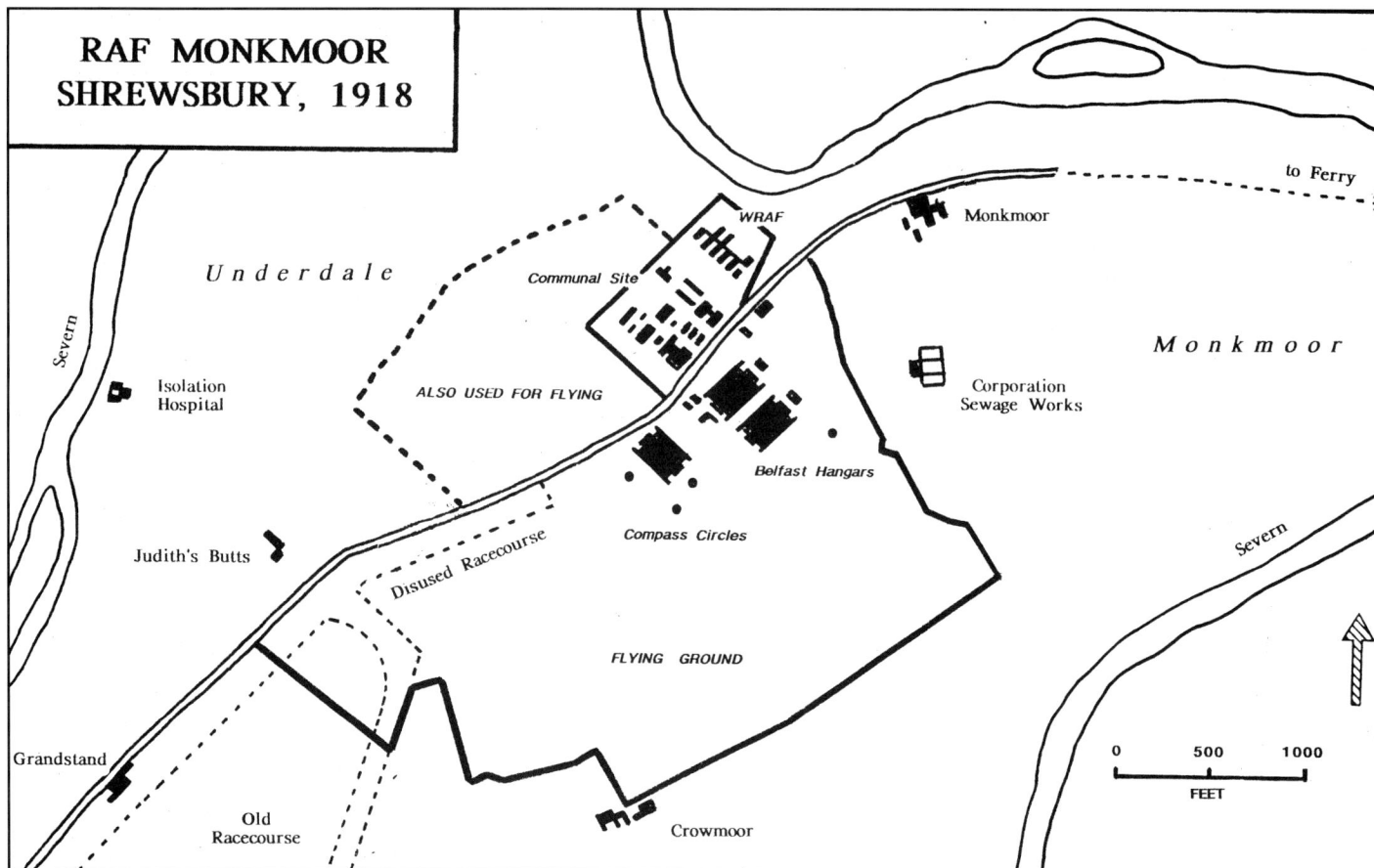

RAF MONKMOOR
SHREWSBURY, 1918

may go some way towards explaining why Tern Hill was always regarded as a 'wet' airfield. It would be October 1941 before work started on concrete runway construction.

Outside Shotwick (or Sealand), therefore, only the tiny airfield at Monkmoor, Shrewsbury, would see civilian flying continue (albeit spasmodically) into the 1920s and early 1930s. As noted earlier, something of a tradition of flying had become associated with Shrewsbury Racecourse and fields adjoining further south along Monkmoor Road towards Abbey Foregate. At its extremes this tradition embraced Charles Graham's ballooning feats of 1838 and Gustav Hamel's Blériot flights of 1912. But whether these slender precedents contributed to the War Office decision to build a First World War airfield 500 yards further north, well outside the built-up area, is open to question.

A large part of Crowmoor (Crowmere) farm and the Old Racecourse (north-east of the Grand Stand and including the beginning of the straight mile added in 1885) was requisitioned in September 1917, the intention being to establish an aircraft storage depot serving 29 and 37 Training Wings. An Air Acceptance Park (later 11 AAP) had been established at minimal expense on 15 September 1917 on Hesketh Park Sands, Southport. It was too remote to efficiently serve the new training depots in Cheshire and Shropshire. Indeed it had come into being specifically to handle new aircraft produced by the Vulcan Motor & Engineering Co. Ltd, Southport who had gained contracts — not as many as first envisaged — for the production of the BE2 fighter, and BE2D/E, DH4 and DH9 and DH9A bombers. This AAP would close on 11 January 1918. In July, following RAF restructuring of its support organisations, AAPs would come on stream at Castle Bromwich (No. 14 AAP) and at Manchester (Didsbury/Alexandra Park, No. 15 AAP). But again, both were geographically distant and both were primarily manufacturers' acceptance grounds, for Avro in Manchester and for Birmingham produced 0/400s and SE.5s at Castle

Bromwich Hence the perceived need for AAP Monkmoor and, as noted earlier, AAP South Shotwick, both unnumbered.

No. 29 Training Wing was formed on 1 June 1917 to control the training aerodromes at Tern Hill and Shawbury which had opened for business in November 1916 and June 1917 respectively. No. 37 Training Wing was officially constituted on 15 October 1917 and had oversight of Hooton Park (operational September 1917) and Shotwick (opened to pupils at the end of October 1917). But in practice, when commissioned in May 1918 under the aegis of 15 (Equipment) Group, Midland Area, with the end of the war (with hindsight) just months away, Monkmoor never rose above a sub-depot to No. 14 AAP, Castle Bromwich. After an active service life of five months, wind down and closure was postponed for a year by the need to handle, store and dispose of the vast surplus of redundant aircraft that poured into the UK from the war zones.

For certain two double Belfast-trussed hangars with tarmac aprons were erected on the eastern side of the track to Monkmoor Farm, with offices, accommodation and secondary airfield on the other side. If the third multiple Belfast hangar, as shown on, accompanying map and working drawing, was built, its subsequent demolition and removal remains shrouded in mystery. Such large hangars were obviously intended for repair purposes and to store dismantled aircraft prior to re-assembly and delivery to the training squadrons. In practice they were not exclusively put to this use as in September 1918 part of the airfield was hi-jacked by 13 (Training) Group with the arrival of Monkmoor's one and only operational unit, the Observer School of Reconnaissance and Aerial Photography (OSRAP), complete with D.H.4s, BE2s, R.E.8s, later D.H.9s, and a couple of Bristol F2b fighters 'for defence purposes'. Whether the unit ever reached its planned establishment of twenty-four aircraft is debatable! Like AAP South Shotwick, operations at Monkmoor were overtaken by the cessation of hostilities. One wonders if the current absence of the fifth and sixth aeroplane sheds signified that they were never built owning to the curtailing of the original plans for the site.

Plans for the aerodrome show a 115 acre flying ground south of, and adjacent to, the Belfast hangars, giving a 1,000 x 500yd landing area. From changes in field boundaries between 1900 and 1925 noted on the respective 6-inch OS plans, it is clear that hedges were grubbed up and the ground levelled — not much of a task with only a six-foot fall west to east. But flying also seems to have taken place from the large field across Monkmoor Road, adjacent to the domestic site.

Monkmoor would appear to have been one of the few airfields that were split by a road and needed simple 'level crossing' procedures and precautions as aircraft taxied from the hangars through the gap in the opposite hedge on to the airfield, *i.e.* a man with a red flag held up such traffic as used this country track which gave access not only to Monkmoor Farm and Hall but also to the much-frequented passenger ferry across the Severn to Uffington.

The increasingly effective use of aerial photography by reconnaissance aircraft along the British front in France and Belgium had prompted the formation of a specialised training unit (OSRAP) whose observers would be saddled with the task of reconnoitering and photographing the whole of the Western Front to a depth of thirty miles behind enemy lines, a dodgy prospect to say the least. The first intake of twelve flight cadets reported to Monkmoor at the end of September for the six week course. Training was completed on 13 November, by which time the normal run of events had been overtaken by the Armistice. During their course the trainee observers photographed the whole of Shropshire and a large part of the Welsh borderland; it is somewhat depressing to think that such a potentially valuable archive has probably been destroyed. The first of the D.H.9As were flown into Monkmoor from Castle Bromwich at dusk on 12 November, but a second intake never materialised. The OSRAP closed on 30 November but a rump rear party was still at Monkmoor in May 1919. After leave the would-be aerial photographers were posted en bloc to No. 2 Fighting School at Marske, near Redcar, Yorkshire, swapping cameras for camera guns, and being demobilised in May 1919, the same time as the latter unit in turn was disbanded and the airfield closed. In view of its frequent appearance throughout Shropshire and Mid-Wales, piloted by B. C. Hucks and John Brereton, it is interesting to note that Marske Sands were the first testing place (in 1910) for Robert Blackburn's monoplane and the original location of his flying school.

The only known flying accident recorded at Monkmoor occurred on Tuesday, 25 February 1919 when both Lt. Charles Edward Peerce, 22, of Hanley Castle, Upton-on-Severn (Worcestershire), and Air Mech. (1st Class) Harry Welch, 20, of Ladywood, Birmingham, were killed. Peerce had been a course instructor with OSRAP and had been retained as a test pilot with the AAP. He had taken off at 11.00 hrs. in an Airco DH9a (F1179) when almost immediately the engine failed and the aircraft spun in from 60ft a quarter of a mile from the aerodrome. Both men were killed instantly. Peerce was buried beneath the wreckage and Welch was still strapped into his seat. As Francis Price Edwards, landlord of the 'Hero of Moultan' inn (on Wyle Cop), rushed to the rescue the 70-gallon petrol tank exploded setting the machine ablaze and scorching the clothing of the would-be rescuer. Edwards had the presence of mind to roll himself in the snow to douse the smouldering threads.

F/Sgt Bowater testified at the inquest that the DH9A had been tested and fuelled up early that morning when 'the engine had been running beautifully'. AM1 Welch had been a mechanic with BSA, Birmingham. Lt. Peerce was a former Yeomanry officer, transferring to the RAF in 1918 where he had become an aerial photography specialist with 2 Squadron, a corps reconnaissance unit, at Hedigneul. He had several narrow squeaks such as that on 28 May 1918 when his Armstrong Whitworth FK8 (C8582) was badly shot up by AA fire over Auchy.

There followed a short period when the airfield was also used for storing surplus Army vehicles, a spill-over from the RAF MT depot up the road at Harlescott, but by August 1919 Monkmoor had been sold. It was officially vacated on 31 October 1919, two weeks after 10 Training Wing (HQ Shotwick) had been formed for the express purpose of closing down the fifteen aerodromes, AAPs, MT depôts, recruiting centres etc. in North Wales and the north-west. One set of hangars and apron were given over to a poultry farm. Hen-houses filled each hangar and birds ranged freely over the hard-standing. Another set was converted into a covered tennis court complex run by James Edward Hince, to such good effect that the 1922 Inter-County Hard Court Tennis Championships were held at Monkmoor.

Across the road a short, gated accommodation crossing had already been laid out. To ease the chronic housing shortage, some of the domestic site buildings were already being re-roofed and converted into the bungalows which today still form the core of the small Glenburn Gardens housing estate. The WRAF 'spider' or linked hutments were converted into a second Monkmoor Isolation Hospital. The first, a tiny building in Underdale Road, was then earmarked for smallpox and cholera patients only. The large flying field was re-fenced, but on a different pattern to pre-war. Crowmoor became a small-holding. A Model Farm was established and the rest of the former airfield shared between seven further small-holdings. As elsewhere in Shropshire this was part of the universal scheme to give 'the boys' (those ex-Servicemen who had survived the horrors of the trenches) their own few acres of English soil. At least such was the theory; in practice the acreages were so small that two of the smallholdings became poultry farms, two the working base for a coal merchant's business, and one the site for a short-lived motor-cycle dirt track! As new housing — Ashley Street, Underdale Road and Avenue — spread up Monkmoor Road, the surviving parts of the Old Racecourse became the Racecourse Recreation Ground.

In the immediate post-war years Monkmoor became the base for Berkshire Aviation Ltd. Here the founders of the company stand in front of their Avro 504K. L–R: Jack Holmes, Alan Cobham, Fred Holmes and R. Graham-Wallard. [Colin Cuddas]

The tradition of flying lingered on in place-names — 'Aerodrome Farm', 'Monkmoor Aerodrome Private Tennis Courts' etc. — right up to March 1940 when 34 MU took over part of the former flying field and the dilapidated hangars, the latter, however, still structurally sound and now housing 'Shropshire Motor Traders', to re-establish a RAF presence within the great loop of the Severn. However, there is also ample evidence to point to Monkmoor being used in the interim by the earliest 'barnstormers', notably one Fred J. Holmes, described in Wilding's *Street Directory* for 1928 as an 'aeronautical engineer' resident at Monkmoor Cottages. This is for the same period as Kelly's *Directory of Shropshire* gives the occupant of Monkmoor aerodrome as Berkshire Aviation Tours Ltd. Actually these were one and the same organisation.

Berkshire Aviation Tours had been formed at East Hanney, between Wantage and Oxford, on 19 May 1919 by Fred and Jack Holmes, literally on a shoe-string — a war-surplus Avro 504K, £50 of aviation fuel, £200 working capital in the bank, and a battered Army-surplus Ford car, the latter essential — until the firm graduated to a lorry — for transporting the portable wind-sock and an advertising banner for putting over the gateway to their rented fields. Second pilot was a young First World War veteran yet to make his name in the aviation world, one Alan Cobham. He, however, would leave the company in the following year.

In the summer of 1919 Berkshire Aviation visited Wolverhampton, Walsall, Kidderminster and Shrewsbury, in fact giving joy rides at 5s. a time (10s. a loop, £1 a spin) from any town with a field big enough to operate an Avro 504. Monkmoor, only recently abandoned, was ideal for the purpose and in 1928 became the north-west base for the company. They had prevailed upon the poultry farmer, one Ernest Arthur Rudd, to lease them one hangar. In 1967 the late Capt. E. E. Fresson, O.B.E., recalled in his book, *Air Roads to the Isles*, how in 1928 he took up a job at Monkmoor as a joy-ride pilot, using le Rhône-engined Avro 504s. His book mentions his surprise at seeing the licensed Ground Engineer, Bert Farminer, deep in the hole he was digging for new gate-posts at the place where 'the planes were moved across the road to the flying-field from the hangars'. 1928 seems to have marked the apogee of the company's fortunes. By then its 'fleet' had grown to four Avros, modified to take two or three passengers. These generally toured the Midlands separately, occasionally coming together for a high-profile display. Serials were G-EAKX, G-EBKB and G-EBKX. Like Alan Cobham after them, publicity was gained by giving the *Shrewsbury Chronicle* and its associated papers such as the *Newport and Market Drayton Advertizer*, tickets for free flights to be won in competition. In return for six column-inches more daring reporters were taken up for free! In their 1919 visit to Monkmoor over 1,000 people paid 6d. just to watch, while thousands more lined the river bank between Castlefields and Ditherington. By 1929 the company had taken up over 55,000 passengers with not one injury.

When exactly flying ceased at Monkmoor is not known, but it might be suggested that it went out with a bang! While the airfield is not marked as such on pre-war aviation charts, there is no doubt that from pre-flight briefings many pilots from the Flying Training Schools and Group Training Pools were familiar with its existence. The large Belfast hangars were still distinctive landmarks from the air. No other reason could explain why, at 18.00 hrs. on Thursday, 16 September 1937, just as the *Shrewsbury Chronicle's* presses were about to roll with the Friday's edition, a Vickers Wellesley medium bomber made a forced landing in Monkmoor. The presses were held and the back page re-made to include a photograph and short report, so short that make, serial and other information vital to the enthusiast is lacking!

According to the report 'a heavy monoplane' in green and yellow camouflage was seen circling Monkmoor and Underdale as if trying to land on the former flying ground. But high tension wires and children playing on the field itself made this impossible. The pilot finally put down on the sports field and recreation ground to the rear of Monkmoor Schools. Its undercarriage collapsed and the aircraft came to rest in a tangle of twisted iron railings. The pilot, P/O Riepenhausen, scrambled out a little dazed, but otherwise unscathed. He had been flying solo from RAF Finningley (near Doncaster) to Sealand on the first leg of a triangular NAVEX and had lost his bearings. In March 1937 'B' Flight, 7 Squadron, at Finningley had received the second production Wellesley for service trials and on 12 April, on reforming as 76 Squadron, was the first unit to be re-equipped with the new aircraft. Police had their work cut out to keep smokers from the wreckage. Oil and petrol had leaked everywhere. A detachment was quickly summoned from Tern Hill to mount guard and oversee recovery.

The other RAF unit in Shrewsbury, the Mechanical Transport (MT) Repair Depôt at Harlescott, survived until March 1932 and was instrumental in encouraging the inter-war ribbon development on Oldheath along the A49 Shrewsbury-Whitchurch road. In 1930 unit strength was some 12 officers and 400 other ranks, exclusive of families in married quarters across the road. For such a comparatively large peace time unit surprisingly little has survived either in the written record or oral tradition. Historians attribute to it the doubling in the population of St. Alkmund's civil parish, from 712 in 1911 to 1,433 in 1921 and 1,811 in 1931. Officer's and airmen's married quarters were erected in 1925 and the children catered for at the tiny Sundorne estate school built in 1849 by A. W. Corbet on the corner of Featherbed Lane in Uffington parish. But in any demographic study it must not be forgotten that hand in hand with the RAF camps came the giant Sentinel Steam Wagon Works (1920) and their pioneer workers' housing scheme in Sentinel Gardens, built to a high standard on Garden City principles, complete with a district heating system of which the long abandoned water tower still stands in Albert Road.

Land between the railway and A49 at Oldheath was requisitioned in November 1917 and Messrs. Parkinsons, the Blackpool contractors, moved in during March 1918. Within the RFC support infra-structure it was originally intended that the Depôt should house an Aircraft Repair Section and a salvage unit. No flying field was attached. In May 1918, shortly after the formation of the RAF and the Area/Group re-organisation associated therewith, these proposed functions were handed over to Monkmoor and individual aerodromes at Hooton, Shotwick (Sealand), Tern Hill and Shawbury, each of which had its own half-Aircraft Repair Section. When commissioned in September 1918 it was intended that the Harlescott Depôt — such was its official designation although Harlescott House and hamlet was some distance away the other side of the railway — should concentrate on MT repair and aero engine overhaul, mainly the Sunbeam Mohawk, Cossack and Maori engines.

To this end No. 7 (North-western) Aircraft Repair Depôt was officially constituted at Harlescott on 1 September 1918, but it hardly got going before it was disbanded sometime in late October, certainly by 1 November 1918. No railway siding was shown on the original plans, but the Depôt was certainly served by one in the post 1922 period, although this may have been shared with W. G. Tarrant's timber

---

RAF records show the Wellesley to have been K7730, delivered to 76 Squadron on 25 June 1937. Remarkably, the wreck was considered salvageable and was returned to makers on 4 January 1938. By the end of June, duly repaired, it was on charge to 27 MU Shawbury before being delivered to 148 Squadron at Stradishall, Suffolk, on 15 September 1938. But this unit had begun 'converting' (if that is the right word for a somewhat retrograde step) to the giant bi-plane Heyford IIIs. Thus it was in May 1939 that K7730 found itself at 36 MU Sealand, the pre-war packing depot, en route to Aboukir Aircraft Depot, Egypt. In October 1939 it was on the strength of 47 Squadron, Khartoum. On 11 June 1940, during the Ethiopian campaign, K7730 was damaged by ground fire and forced down in Italian-held territory. Its three crew became prisoners of war.

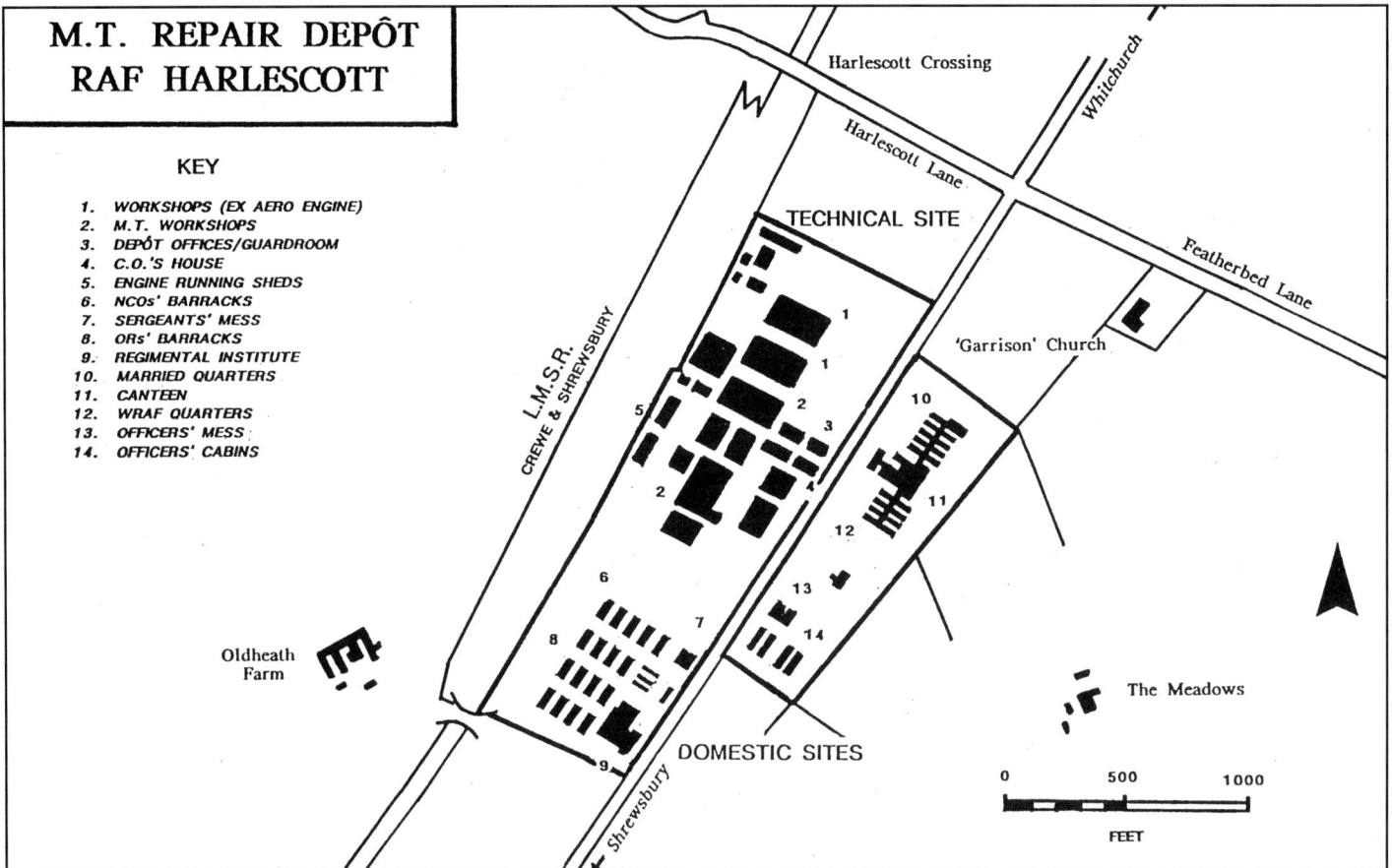

M.T. REPAIR DEPÔT RAF HARLESCOTT

KEY

1. WORKSHOPS (EX AERO ENGINE)
2. M.T. WORKSHOPS
3. DEPÔT OFFICES/GUARDROOM
4. C.O.'S HOUSE
5. ENGINE RUNNING SHEDS
6. NCOs' BARRACKS
7. SERGEANTS' MESS
8. ORs' BARRACKS
9. REGIMENTAL INSTITUTE
10. MARRIED QUARTERS
11. CANTEEN
12. WRAF QUARTERS
13. OFFICERS' MESS
14. OFFICERS' CABINS

yard and mills which later occupied an empty corner of the camp. The existence of a narrow gauge internal tramway linking all workshops presupposes an exchange siding with adjacent LNWR tracks. A piece of the tramway still remains embedded in concrete in the foundations of one of the demolished GS sheds. The tramway was lifted *c*.1921 as mobile lifting gear — hoists and the ubiquitous Ransome & Marles crane (with its solid rubber tyres) — supplemented sheerlegs and gantries.

The Depôt was not fully operational until November 1918 and local tradition has it that in the post-Armistice 'detente' skilled German POWs helped in the installation of machinery in the MT workshops. Indeed, the designation No. 2 MT (POW) Repair Depôt that occasionally crops up in official records (strangely of RNAS provenance) would seem to suggest that suitably qualified POWs may also have been employed on the shop floor until such time as they were repatriated and replaced by the service's own technicians as they returned from abroad. The production and overhaul of aero engines ceased abruptly at the end of hostilities and the Depôt became a temporary dumping ground for surplus vehicles of disbanded units, overspilling at times onto Monkmoor aerodrome.

It was the processing of thousands of these vehicles for the commercial market that enabled the Depôt to survive post-war threats of closure. In June 1919 the Harlescott unit became the RAF's main MT Repair Depôt under the command of S/Ldr. R. H. Berney OBE He was succeeded in 1923 by W/Cdr. R. J. Mounsey, OBE, a First World War veteran who, in May 1916 as a raw 2/Lt, had received a rude introduction to the air war over the Western Front flying a BE2c on artillery registration sorties with No. 2 Squadron based at Merville and St. Omer. The Depôt's last CO was W/Cdr. R. F. S. Morton, former RFC major, and OC 11 Squadron from August 1917 until 17 May 1918 when he was wounded in action flying a Bristol FE2b out of Remaisnil.

In addition to overhauling RAF vehicles and painting them in the appropriate colours for operations overseas, the Depôt converted the Fordson agricultural tractor for fuel bowser and aircraft towing (replacing the spiked wheels with heavy-duty rubber tyres), built Leyland fire tenders, towing trucks and ambulances, and from the early 1920s the famous and much revered Hucks starter, built on a Model T Ford chassis, which was a familiar sight on aerodromes until the mid-1930s. The depôt's closure was announced in September 1931; existing contracts were to be completed with work finally ceasing on 31 January 1932. Personnel were posted out, mainly to the Home Aircraft Depôt (former Inland Area Aircraft Depôt) at Henlow (Bedfordshire). This was a service manned Aircraft Repair Depot, including an Engine Repair Section, and in 1938 would become No. 13 Maintenance Unit. Workshops were stripped and Harlescott finally shut its gates on 31 March 1932. The flimsier buildings were demolished but the concrete sheds were rented to small industrial tenants.

Three of the four giant workshops, coupled general service sheds with Belfast truss roofs, yet survive within a perimeter fence as do a couple, of singleton sheds around the former MT square. The main entrance and drive was by Kendal Road with SCR Retail occupying the Depôt's former administration block and the CO's house (hut) being swallowed up by Wickes (builder's supplies) store and yard. Sunlight Workwear Services occupy some former storage buildings alongside the railway, but the north end of the camp has disappeared under a massive BT complex and a Honda dealership. The former communal area at the south end of the camp was sold in 1934 to Fletcher Homes Ltd., a Blackpool development company, and has long given way to private housing — Windermere, Grasmere and Coniston Roads — typical semi-detached pairs of the 1930s.

The former married quarters across the road have been largely lost to Meadow Farm Drive, Harlescott Close etc. , although in the case of the latter road, as at 'Glenfern' at Monkmoor, the architecture and positioning of houses proclaim up-graded and modified RAF buildings

The Main Gate and Guardroom, RAF Harlescott, now Kendal Road. [SRRC]

Above and below: Some of Monkmoor's former aeroplane sheds are in a better state of preservation than others.

Above: The former Guardroom building at RAF Harlescott as it is today.

Left: The distinctive profiles of the much-modified aeroplane sheds are still a feature of the approach to Monkmoor Industrial Estate.

still extant over seventy years later. The loss of most of the Harlescott facility to civilian use was facilitated by the fact that also In 1934 St. Alkmund's Civil Parish was dissolved and incorporated largely into Shrewsbury Municipal Borough, which authority moved rapidly to remove or develop the semi-derelict military hutments.

In 1938/9 industrial tenants were evicted from the former work shops as the War Department took over the rump of the former MT Depot, again handling vehicles, but serving also as a trans-shipment and packing point for war materials produced by other factories along the Whitchurch Road. Sentinel (Shrewsbury) Ltd. produced Bren-gun carriers (400 a month by 1943), thousands of 3. 7 and 4. 5-inch shells, refrigeration units, D-day tank landing ramps etc. all of which was trans-shipped from the Harlescott depôt, as were the crated wing sections from the 'Spitfire factory' at the rear of the Midland Red bus garage in Ditherington. After the war Harlescott became for a short while a REME Armoured Fighting Vehicle Depôt (AFVD), storing and disposing of armoured cars, half-tracks etc. The whole complex was then sold to the Ministry of Agriculture for storage purposes. Today brand new buildings alongside the railway house the Ministry of Agriculture, Food & Fisheries, West Midlands Division, Veterinary Investigation Centre. The remainder of the First World War sheds are empty and, vandals permitting, up for sale.

With the outbreak of war in 1914 private and commercial flying had ground to a halt. Five years later the aviation scene had changed out of all recognition. At the Armistice in 1918 there were some 301 aerodromes and/or landing grounds in Britain, 256 of which would be de-requisitioned over the next two years. Additionally hundred of military aircraft would become surplus to peace-time requirements, along with their redundant pilots. To capitalise on this largesse crash legislation was drafted to regulate civil and commercial flying from its re inauguration on 1 May 1919.

But the proposals, when published on 25 April 1919, proved to be disappointing for North Wales and the border regions. No commercial services had been approved or licences granted to local companies. A tentative prospectus had been issued by the Liverpool-based Great Northern Aerial Co. to provide scheduled services between Liverpool, Dublin, the Isle of Man, and the North Wales coastal strip Rhyl-Colwyn Bay-Llandudno-Conwy-Holyhead, with perhaps even a Holyhead-Dublin service, but these plans foundered as backers got cold feet. The Government made one concession only to Welsh geography. Arrangements were pencilled in for emergency landing grounds along the Irish Mail route, Sealand-Holyhead-Dublin, should it ever materialise. It did not prove feasible and was never implemented. Such designs merely serve to underscore the current fragility of both men and machines when it came to spanning large stretches of open water. As will be seen later, the political and military aspects of the 'Irish connection' would loom large in the early history of RAF Shotwick.

But at this time, given the primitive navigational aids, the lack of radio and the absence of an air-sea rescue service, the Irish Sea was a formidable obstacle to aviators. Nowhere is this better illustrated than by reference to the fiasco and near-tragedy that passed for the first post-war trans-Atlantic crossing attempt on 18 April 1919. Five years earlier Gustav Hamel had been a serious contender with a specially modified Martin-Handasyde monoplane. His disappearance over the North Sea in May 1914, followed quickly by the outbreak of war, temporarily signified the end of such frivolities. With the cessation of hostilities the attention of the aviation world turned once more to the last great challenge. The £10,000 *Daily Mail* prize for the first trans-Atlantic crossing was still to be won.

A Major Wood and a Captain Wyllie, two former RAF officers, were first off the mark. It was not a military venture, but they had the support and sponsorship of Short Bros. of Rochester, Kent, the oldest established firm of aeroplane designers and manufacturers in the United Kingdom. From 1911 onwards Shorts had particular success with the design of seaplanes. It was a Short seaplane that first carried a torpedo aloft and it was in a Short seaplane that Flt/Commander C. H. K. Edmonds made history by torpedoing a Turkish transport in the Sea of Marmara.

Just before the Armistice the firm had produced the Short N2b, a bomber float seaplane. Only two were constructed (serials N66, N67), the rest of the order, like so many others in the pipeline when the war ended, being abruptly cancelled. At the same time they brought out the Short 'Shirl', a torpedo-carrier biplane. This was generally similar to the N2b save that it was fitted with a divided wheel and skid chassis for flying off decks and to allow the stowage of an 18-inch torpedo under the fuselage. Three 'Shirls' were constructed initially, the order for a further 100 (serials N7550-N7649), placed with Blackburn Aeroplanes, also being cancelled. It was one of the three (N111 *Shamrock*), given slightly larger wings and tailplane and with a massive 300-gallon long-range fuel tank slung torpedo-fashion under the belly, that Shorts prepared at their Eastchurch works for the ill-fated trans-Atlantic attempt. Possibly, after seeing so much design work and effort gone down the drain, the company sought to recoup something in the way of prestige and market generating publicity, that is, if the project could be successfully carried through! But, perversely and against all odds, Wood and Wyllie elected to make their trans Atlantic attempt flying east-west against the prevailing westerlies. At that time of year the latter were re-inforced by Polar-maritime winds and intense cyclonic disturbances or depressions, but with meteorology in its infancy these were as yet unquantifiable and little appreciated hazards. To further lengthen the odds the chosen starting point was The Curragh, Kildare, some 100 miles from the western extremities of Ireland. The first task was to reach there. Holyhead was a vital staging point or navigational landmark before turning west across some 80 miles of sea. Throughout the early stages Shamrock was escorted by John Lankester Parker, Short's test pilot, in a second 'Shirl' (serial N112). The *Holyhead Chronicle* would devote but a few scathing column-inches to the events that would be played out just beyond the harbour…

The trio left Eastchurch at 15.15 hrs. on the Good Friday (possibly not the best of omens) and departed Holyhead at about 19.20 hrs. Twelve miles out to sea Shamrock's engine failed (an airlock in the fuel feed system) and it turned round heading for the Anglesey coast. It fell short and landed in the sea half a mile outside the Holyhead breakwater. Wood was rescued by dinghy and Wyllie by the Holyhead lifeboat The latter also towed *Shamrock*, floating docilely upside down under its great pseudo-buoyancy tank, into harbour. Strange to relate, in the excitement on the harbour front nobody seems to have remarked on the fact that J. L. Parker in N112 also came to grief. Seeing what had befallen *Shamrock*, he had attempted to land in a field just outside Holyhead, but over-ran and crashed into a stone wall. Fortunately he was uninjured. So ended this particular trans-Atlantic attempt. The two

After the debacle of the trans-Atlantic attempt of April 1919, the Short 'Shirl' 2b bomber N112 was rebuilt as a land plane with a mail container suspended under the fuselage. [Raymond Davies]

'Shirls' were recovered and taken back to Eastchurch where N112 was rebuilt and converted as a two-seater. Later photographs show it with a mail-container slung underneath the fuselage where once was suspended a torpedo or long-rage fuel tank. Wood and Wyllie fade from centre stage, but others were queuing in the wings to make their entrance.

On the evening of 14 June 1919 John Alcock and Arthur Brown took off from Newfoundland in a converted Vickers Vimy IV bomber. Seventeen hours later they landed within sight of the masts of Clifden radio station, Connemara, the first in a long line of pioneer aviators to hog the trans-Atlantic limelight. The ill-conceived attempt of Wood and Wyllie, if recalled at all, is usually in the context of the anecdotal trivia of the Holyhead-Dublin 'Irish connection'.

It must be admitted that in drawing up his blue-print Sir Frederick Sykes, first Controller of Civil Aviation, showed a better grasp of the geography and economic potential (or lack of it) of the Principality of Wales than many of its inhabitants. It would be another forty years before regional planning attempted to break out of the 'industrial coffin' and 'megalopolis' concepts beloved of geographers, which left Wales and the border very much peripheral to the London-Midlands-Merseyside industrial axis. Matters were exacerbated by (a) the paucity of suitable existing airfields that could be taken into service — Wales had not been over-endowed with war-time military installations — and (b) the understandable reluctance of landowners to part with the 70-80 acres considered necessary for a commercial airfield starting from scratch.

It took some time for the inhabitants North Wales to appreciate the negative provisions of the Air Navigation Bill as there were events still taking place along the coast that kept aviation to the forefront and coloured the ambition of forward thinking local authorities. On 7 July 1919 — a far cry from the relatively tiny blimps of RNAS Llangefni that regularly used the playing fields of Llandudno as emergency landing grounds — the great rigid airship R33 cruised slowly at 25 knots along the North Wales coast in support of the Victory Loan Fund. It was a period when the nation was being reminded that the cost of war was not to be reckoned solely in terms of human lives lost. While the airship SR-1 had earlier made a 25 hour promotional flight distributing Victory Loan leaflets over South Wales and the West Country on 6/7 July, the R33 made a corresponding flight of some 900

miles in 31 hours over North Wales, the north-west, and parts of Ireland.

R33 was Britain's most successful rigid airship, 643ft long, 76ft in diameter, powered by five 240 h.p. Sunbeam Maori 4 engines, and with an envelope capacity of 1,950,000 cu. ft. She was an almost identical copy of the Zeppelin L.33 and drew enthusiastic crowds wherever she appeared and, presumably, a fervent patriotic response in the purchase of Victory Bonds. After dropping leaflets with remarkable precision over Liverpool Pier Head, Water Street and Dale Street, the R33 crossed the Mersey to Birkenhead and on to Rhyl and Llandudno before arriving at Bangor 'at an opportune moment — just when the mayor was addressing a very large crowd by the town clock on behalf of the Victory Loan Campaign', the local hack little realising that some careful planning had gone into bringing this 'coincidence' about. Considerable care had to be exercised in dropping leaflets — even a wad of paper could do considerable damage to the air screws. Duty done the R33, under the command of Capt. Thomas, cruised the Menai Straits at 20 knots — the newspaper report does not mention a brass band playing on the upper gun platform — before turning westwards towards Holyhead and setting course for Dublin. The R33 arrived back at RNAS Pulham (Norfolk) late on 9 July having 'plastered' Manchester and its satellites en route.

But even before the provisions of the Air Navigation Act were known Rhyl Urban District Council sought to secure a piece of the aviation action, no doubt recalling the impact made on local tourism by the pre-war aerial displays by Vivian Hewitt in his Blériot XI. In 1919 the UDC may have succumbed to sales talk by A.V. Roe & Co., the Manchester aircraft builders who, in the immediate post-war period, were faced with cancelled orders and massive redundancies. The war had boosted public interest in flying and a perceived market for pleasure flights or 'joy rides' seemed an ideal means of generating new money for the company. In advance of the Air Navigation Bill special clearance was obtained to fly the public over the Easter period from the beach at Rhyl, using war surplus Avro 504Ks converted to three-seaters and flown by ex-wartime pilots. The trial week-end was a great success — people were just starting to holiday again after the deprivation of war. Thus was ushered in the age of the 'five-bob flip' — save that in 1919 it cost a guinea or more!

The Avro Transport Company was formed to co-ordinate pleasure flights nation wide and to run a scheduled air service between Manchester–Blackpool–Southport. In May, John Lord of Avro attended a meeting of Rhyl UDC and obtained permission to use the East Sands as an aerodrome/flying field for pleasure flights. Possibly the Council were looking for something more substantial and prestigious in the way of a Manchester-Rhyl scheduled feeder service but this was not considered financially viable. Government approval for an aerodrome was quickly given, no doubt seeing it as a potential emergency landing ground should any scheduled service to Ireland via Holyhead materialise.

After the trial run at Easter, the Avro Transport Company were back in Rhyl for the Whit week-end (7 June 1919) 'when a large section of beach was flagged off and three of the latest Avro machines fitted with le Rhône engines were lined up on a space on the aerodrome opposite Plas Tirion Terrace. Discharged soldiers were engaged to keep the runaways clear of the public. Major MacInnes, Capt. Heriot and Lt. Hudson, officers who had

In July 1919 the rigid airship R33 cruised the North Wales coastal resorts to promote the Victory Loan Fund.

Avro Transport Ltd., was a pioneer of pleasure flying from the beaches of North Wales and Lancashire. A scene typical of Rhyl (East Sands) and Kinmel during the summers of 1919–20. [Raymond Davies]

served with the RAF, were the pilots'. While people flocked to see the aeroplanes by day there was also some trespassing at night by courting couples who resented their traditional mating grounds being roped off and their lustful activities curtailed! Flares fired from Verey pistols soon ended this nuisance.

Later a group of officials from Rhyl were shown round Alexandra Park, Manchester and then were flown home. They were evidently still pressing for a scheduled air service from and to the town — local councillors needed little convincing of 'the pre eminence of aviation' and the role Rhyl should play in civil and commercial flying. However, back on East Sands pleasure flights were flown daily between 10am–8 pm for the next seventeen weeks without a forced landing or accident, although occasionally flights were cancelled because of bad weather. A regular visitor was J. C. C. Taylor, Avro's Chief Engineer and trouble-shooter, in his one-off Avro 538 (temp. reg. K132, later G-EACR), that had started life as a private venture Home Defence 'Scout' machine derived, in design and probably parts, from the Avro 531 'Spider'. With excellent manoeuvrability and a 150 h.p. Bentley BR. 1 engine it may have gone into production had not the German air attacks begun to peter out.

The 'joy rides' from East Sands were quoted as: 'Short Circuit (over Rhyl, Foryd and Prestatyn) £1 1s. 0d.; St Asaph/Rhuddlan–Bodelwyddan/Abergele £2 2s. 0d.; Looping the Loop £2 2s. 0d.' The booking office attracted its own crowds to watch customers book for their guinea 'flip' or two-guinea 'aerobatic'. Saturday night was 'fun night' as Welsh-speaking lads from 'up the Vale' hotly debated the merits of a flight to finish off the rounds of Rhyl's flesh-pots. To see them being helped into their flying kit whilst trying to comprehend instructions in a 'foreign' tongue was almost a pantomime in itself. The record for pleasure flying must have gone to Col. Malcolm A. Colquhoun, Commanding Officer of the heterogenous

units of the CEF at Kinmel Camp. The unpleasantness and notoriety of the March mutiny was behind him. He was billeted in Rhyl and had over twenty flights, sampling every known 'stunt', and even booking a plane for an hour's flight or more. Frustratingly he would hint to disinterested reporters at the use made of aircraft by the army in the area. In the light of the earlier mutiny one wonders whether aircraft were now being used to speed communications between the camp, the embarkation ports of Liverpool, Glasgow and Avonmouth, and OMFC HQ, London or whether they were being used for 'Army co-operation' in the rather pointless training programmes that had contributed to the accumulative grievances that had triggered off the mutiny.

In August 1919 the Chairman of Rhyl UDC concluded his mid-term review of the successful flying season from East Sands with the words 'We will be in the forefront of commercial flying on the North Wales coast'. His prophecy has a hollow ring when headlines such as 'SUCCESSFUL INAUGURATION/COUNCILLORS AND OFFICIALS GO ALOFT' conflicted with 'BREAD QUEUES AT RHYL' symptomatic of post-war ills and blues.

In May the following year the Avro Transport Company announced that, following the Company's success at Rhyl, aviation was to commence on Morfa Conwy or Conwy Marsh, the sand dune spit on the left-hand bank of the Conwy estuary, traditional site of Territorial and Yeomanry summer camps. During Whit week Saturday, 23 May-29 May 1920, two-guinea flights would be made to Betws-y-coed and over Bangor, the Menai Straits and Puffin Island. But this particular venture, if it got started at all, appears to have been abruptly terminated. During the 1919 season ATC machines on Rhyl's East Sands were never so busy as when a large number of charabancs had pulled into town. Many men came expressly to have a 'flip', indeed, as the local press had it, 'men who preferred a muffler to a collar readily pay the extra guinea for an 'aerobatic'. They want thrills!'

But prospects for the 1920 season were not propitious. People who in 1919 had flocked to fly as an expression of post-war delirium and release were finding that jobs and cash were now much harder to come by. Despite numbers carried in the 1919 season the ATC was not even paying its way. All it appeared to be doing was fostering air-mindedness, an expensive bit of philanthropy! In April 1920, after a poor Easter, which fell early on 4 April, the Avro Transport Company was wound up. Pleasure flights along the North Wales ceased abruptly. G-EACR was lost to Flintshire skies, being sold on to Canada. Several aircraft were acquired by individual pilots and smaller concerns such as Berkshire Aviation to carry on the financially precarious barnstorming tours that would be so much a feature of local aviation in the 1920s and early 1930s.

On a broader front, during the Railway Strike of 1919, Wrexham, like other large towns, made full use of the *ad hoc* Government-organised airmail service to get urgent business mail to and from London and other regional centres. There was a limited local mail service by road, but London mail took three to four days in transit, either travelling with the milk convoys to Manchester, one of the northern termini of the provisional airmail service, or by rail to Welsh Road Halt (later Sealand Station) which lay in between the North and South Camps, RAF Shotwick on the GCR Hawarden Bridge–Chester line, which aerodrome provided a feeder link between Manchester in one direction and Castle Bromwich and London in the other, using war-surplus D.H.9s and recently demobilised pilots. Despite pickets, volunteers were managing to run two trains daily along the Wrexham–Seacombe line.

Such stir and flurry was concentrated into some three weeks during which Shotwick's erstwhile AAP saw more aerial activity than it ever did in the last months of the war. From here mail was distributed throughout North Wales, Wirral and Merseyside by aircraft of the 'Shotwick Circuit', with aircraft landing in any convenient field, sometimes spending the night away before returning to Shotwick with outgoing mail. To this period is owed the emergence of the *Aeroplane Field* in Hightown, Wrexham, safely under the walls of the RWF barracks!

By mid-October the strike was over. Thereafter, apart from a few private owners operating from landing-strips on their estates, and a handful of barnstormers, there would be little civilian flying in North Wales and Shropshire in the 1920s. Aerial activity would remain largely the province of No. 5 Flying Training School at Shotwick (Sealand). All of which makes it easier to comprehend the relative lack of success pursuant upon Sir Alan Cobham's Municipal Aerodrome Campaign of 1929 and the National Aviation Day Displays of the next four years, when, with the active support of the Air Ministry and the Air League of the British Empire, he struggled to improve the air-mindedness of the people of Bridgnorth, Shrewsbury, Whitchurch, Oswestry, Wrexham, Bangor, Rhyl, Beaumaris, Pwllheli, Aberystwyth and all points west! Apart from pampering to municipal egos, he must surely have realised he was flogging something of a dead horse!

National and local headlines

Although in the immediate post-war years British aviation was in the doldrums men from the North Wales and Shropshire, especially former members of the RFC and RNAS, did participate in civilian flying elsewhere. Frequently they made the headlines. One such was Howard Saint, DSC, ex-RNAS, who, in September 1919, won the big flying race around Holland in competition with over forty other aviators from around the world. Carrying two passengers, his machine reached speeds of 153 m.p.h.. He had been in Amsterdam six weeks and was moving on to Athens on the next leg of the 'Grand Prix' circuit which had quickly emerged from the chaos of war.

F/Cdr. (later Capt.) Howard John Thomas Saint, one of Wrexham's unsung air 'aces' of the First World War, was born on 21 January 1893 at Ruabon, Denbighshire, the son of T. W. Saint, mining engineer working in the collieries of the Ruabon Coalfield. 'Unsung' perhaps because the British never consciously promoted the 'ace' image. This term was first used by an Allied journalist in reference to the French pilot Roland Garos who shot down three enemy aircraft and forced two others to land in a period of eighteen days. In the news stories it was reported that five 'scores' were required to become an 'ace'. This misrepresentation was taken up and quickly became an accepted criterion for 'ace' status.

Howard Saint served 1915–16 in France as a Chief Petty Officer in the Royal Naval Armoured Cars. Commissioned in August 1915, he was seconded for pilot training, gaining his wings on 2 September 1916 and being posted 8 Flt. B Sqn. 5 Wing (later 5 (Naval) Squadron), a RNAS bomber force based at Coudekerque, near Dunkirk, flying Sopwith 1½ Strutters. In October naval air units also started to move south to assist the RFC on the Western Front. Now with 10 (Naval) Squadron at Droglandt, Saint started scoring, appalling weather notwithstanding, during the great ground struggle that was the Third Battle of Ypres and the fight for the Menin road, when the RNAS and RFC were called upon to sustain ground attacks and give low-level support for infantry. On 16 August 1917, whilst flying a Sopwith Triplane. he was wounded in the side by machine-gun fire from the ground, but it could not have been serious as he was in the air again five days later. His 'score-sheet' for 1917 reads as follows:

1917	Aircraft	Location	Enemy Aircraft
9 Aug.	N5380 Triplane	Polygon Wood	Albatros III — out of control
14 Aug.	N5380 Triplane	Houthulst Forest	Albatros III — in flames
16 Aug.	Sopwith Triplane — wounded		
21 Aug.	N6295 Triplane	s. of Roulers	Albatros C — out of control
25 Aug.	N6295 Triplane	s. of Roulers	Albatros D.V — out of control
21 Sept.	B6201 Camel	Wervicq	'Scout' — out of control
23 Sept.	N6341 Camel	Westroosebeke	Albatros D.V — out of control
20 Oct.	N6341 Camel	Dixmude	Albatros D.V — destroyed

Saint left 10 (N) Squadron on 14 November for Home Establishment and later that month was awarded a well-earned DSC. After his stint around the racing circuits of Europe, he became Chief Test Pilot for the Gloster Aircraft Company, being particularly involved with the development of the Gloster Gamecock (1925/26) and 'experimentals' and prototypes such as the Gloster Guan (August 1926), the Goldfinch (May 1927), the Gloster Gambet for the Japanese navy (December 1927) and the metal-structured SS.18/SS.19 (1929/30). He flew the prototype Hawker Hornbill at the 1926 Hendon Display. Latterly he tested the GAC's ventures into multi-engined aircraft design, the Gloster AS.31 Survey (G-AADO in June 1929) and in February 1932 the GAC's 'Jumbo', the TC.33 bomber/transport (J9832). On 1 December 1933, whilst flying an experimental version of the Breda 15 (G-ABCC) it developed wing flutter and crashed on Churchdown Hill. Saint was unhurt, but the Breda was written off. In 1934 he was replaced as GAC's Chief Test Pilot by P. E. G. Sayer, an erstwhile assistant test pilot at Hawker. But this was not quite the end for Saint. In October 1934 he was busy test-flying the Parnall G.4/31 at Yate (Gloucs.), a former First World War aerodrome where George Parnall & Co. were building and testing a number of experimental aircraft of their own design. But in 1935 Parnall sold out. The main interest of the new company lay in power-operated gun turrets — and so Capt. H. J. T. Saint fades from the aviation scene.

Occasionally it would be the full-scale military funeral in a quiet country churchyard that would inform the local community that one of their own had been actively involved, however humble the role, in prosecuting the interests of Britain and Empire in the air. Thus on Monday, 23 March 1931, some five weeks after he was killed, 19-year old LAC William George Steven, RAF, was laid to rest in St. Paul's Churchyard, Isycoed, near Wrexham. On top of the flag-draped coffin were just two wreaths — from his widowed mother and the RAF squadron that he had joined less than six months previously. The honour guard and firing party were drawn from 5 FTS Sealand (as the nearest RAF station) and from RAF Halton (Buckinghamshire), home to No. 1 School of Technical Training, where Steven had very recently been an apprentice, one of 'Trenchard's Brats'. They were under the command of F/Lt. Lascelles of RAF Mount Batten, (Plymouth). Was it a coincidence that scarcely had the forlorn notes of the 'Last Post' and 'Reveille' died away over the Dee meadows than three D.H.Gypsy Moths from Sealand trundled into view? Or did they just happen to be on a routine flight, making for the church at Bangor-on-Dee, a familiar land mark and turning-point for trainee pilots on a NAVEX?

It was a big funeral. Steven lived at Bowling Bank Farm and had attended Isycoed V. P. School before obtaining a bursary to Grove Park Boys Grammar School, Wrexham, which he left in 1927 having passed the Halton entry examination (equivalent to School Certificate). LAC Steven had served with 209 Squadron ever since completing his three-year trade training. The unit had re-formed at Mount Batten (formerly RNAS Cattewater), Plymouth, in March 1930 specifically to operate the new Blackburn Iris Mk.III flying boats. Over 67ft long, with a wingspan of 97ft and powered by three Rolls-Royce Condor IIIBs, these giant bi-planes were from 1930-34 the largest aircraft in the RAF. Only four were built and squadron strength was made up with the Supermarine Southampton and other ancient boats. Steven had enlisted for 12 years. Trade vacancies in various RAF units were filled by Halton apprentices in a pecking order based on final examination success. A posting into flying boats with LAC rank and great prospects must have seemed the fulfilment of all a young man's ambitions. Little did anyone think that within six months these dreams would be rudely shattered.

209 Squadron took delivery of its first Iris late in February 1930, the service prototype, N238 which for three months had been undergoing trials. This was powered by Condor IIIAs and had first flown on 21 November 1929. It was intended that the Iris IIIs should largely function in a reconnaissance role for the Fleet, but like its sister aircraft N238 also undertook long-distance experimental-cum-'showing the flag' flights. On 20 January 1930 N238 set off for the Persian Gulf to assess, inter alia, the effects of operating a large flying boat under tropical conditions, especially at take-off. But after reaching Malta on 23 January the project was aborted, N238 returning to Mount Batten on 14 February. N238 hit the national headlines when, on 4 February 1931 whilst making a final approach to moorings in squally weather, her nose suddenly dropped and she flew into the water at Mount Batten at over 70 m.p.h. and disintegrated, losing wings and engines and the hull sinking like a stone in deep water. Nine people on board were killed. Eight were trapped inside the hull, including W. G. Steven. There were four survivors, but one died later in hospital. The accident threw a gloom over the whole of the Wrexham area, accentuated by the problems and long delay in salvaging the Iris's hull and recovering the bodies. Since the normal complement of an Iris III was five air-crew, the high casualty figure would appear to suggest that the last flight of N238 was either a domestic communications one or an air test/experiment, carrying technical officers from the MAEE (Marine Aircraft Experimental Establishment) and extra fitters and riggers.

That the local populace could empathise thus with the loss of the Iris III, can be explained by the fact that flying boats were no strangers to the inhabitants of north-east Wales, especially those of the coastal strip. Thousands had seen these huge machines, yet, tantilisingly, despite careful and detailed planning, few had set foot in one, a sad state of affairs attributable directly to the unpredictable weather which seemed to plague RAF public relations flights to North Wales.

Widely billed for 12 September as a great end of season attraction at Llandudno, three Supermarine Southampton I flying boats (N9896, S1036, S1037) of 480 (Coastal Reconnaissance) Flight based at Calshot, did not even attempt to land, but just circled the resort once before disappearing in the direction whence they had come! 'Risk of collision on take-off' was the official excuse for the fiasco, but this cut no ice with irate councillors who considered that the 'goodwill visit', far from cementing public relations, 'has made a fool of the whole of North Wales'. A motion to write to the Secretary of State for Air saying that 'if his machines cannot land in Llandudno the things are no good' was tactfully dropped on a point of order! The official excuse was not as facile as it sounds, since the three flying boats were 480 Flight's only serviceable machines at the time and had to be carefully nursed! As it was, numbers would be reduced to two (S1036 & S1037) for a visit to Rhyl scheduled for Saturday, 17 September 1927, possibly to compensate for the fiasco of the previous week. 480 Flight, one of the few post-war coastal units, had been formed on 1 April 1922 from the rump of 230 Squadron. It was attached to the Home Fleet and exercised with them, but once a year went on a PR 'summer cruise' around the coast of the UK or further afield to the Baltic and Mediterranean. Southhamptons had been taken on strength in August 1925, so they were as yet still something of a novelty to the lay observer. The unit achieved squadron (No. 201) status on 1 January 1929.

It was intended that after a short display over Rhyl the Southamptons should land in the sea off Plas-dŵr and taxi into the more sheltered waters of Foryd harbour, where they would be visited by the chairman of the UDC, other local dignitaries, and privileged permit holders. Large crowds on the Esplanade and Foryd wharf watched the huge biplanes arrive dead on schedule and circle the town, passing just a few feet over the dome of the Pavilion, their Napier Lion engines deafening the assembled throng. At 52ft long, 22ft high, and with a wing span of 75 ft., they made an impressive sight. But only one landed (S1037), some 150 yards out and in choppy seas. It started its taxi-run towards the Foryd, but obviously its crew had second thoughts and immediately took off again. It circled Rhyl once again before rejoining its companion and bearing off towards Colwyn Bay and the Great Orme. It was 'all very exciting' according to a local reporter caught in a strangely benevolent mood. His valediction only hints at the great social occasion missed: 'As they passed over the West Parade the men in the flying boat were waving their hands to the girls on the promenade. The ladies responded by waving their handkerchiefs'. A telegram was later received from the flight commander apologising for having to abort his mission and stating that sea conditions prevented the boats from anchoring off the town.

Lay people perhaps did not realise that the short history of flying boats was cluttered up with beachings, taxying collisions with sea walls, buoys or boats, driftings from moorings, loss of floats, the holing of hulls, sinkings during gales etc. in addition to the everyday concomitants of early flying such as engine failure, running out of fuel and getting lost! Caution was justified when one considers the eventual loss (in Welsh waters) of Southampton S1036 which, on Saturday, 30 June 1928, en route from Calshot to Barrow-in-Furness for anti-submarine exercises, was forced down with engine trouble into heavy seas off Bardsey Island. Fortunately, after a tense four hours the crew of five were with great difficulty picked up by a Holyhead trawler — just as the flying boat was on the point of sinking.

Following the coast of Cardigan Bay, hopping over the Llŷn

peninsula into Liverpool Bay, thence taking leave of Point of Ayr (Flintshire) for the Isle of Man and points beyond, was a recognised seaplane route between the Solent, north-west England and western Scotland. Inevitably accidents to flying-boats were just waiting to happen somewhere along the North Wales coast. Thus on Friday, 24 February 1933, three Southampton IIs S1230, S1234 and S1235 of the Seaplane Training Squadron, Calshot, ran into a snow storm driven by gale force winds as they approached Rhyl on a flight from Stranraer to base. After making little headway in almost zero visibility two of them, S1230 and S1235, sought refuge in Yr Hen Borth ('The Old Harbour'), Porthdinllaen. It was a case of 'out of the frying pan into the fire'; rather than risk losing their fragile moorings and being swept out to sea by the intensifying storm, the aircraft were deliberately beached, sustaining considerable damage, especially S1230 which ended up athwart a sea wall, its starboard wing submerged and folded back. However, both were salvaged and rebuilt. S1235 was back in service with the STS in October 1934 and S1230 by April 1935.

Flying-boats apart, the empty moribund skies over Liverpool Bay were occasionally graced by the passage of Britain's new breed of rigid airships as they were 'run in'. Five times longer than the familiar Llangefni blimps of yesteryear and anything up to 78 times the cubic capacity, these giants of the sky excited enormous local interest, with their subsequent comings and goings charted with almost proprietary zeal by Flintshire and Cheshire newspapers. Seventy years on their majestic, lumbering transit still remains one of the indelible highlights of childhood memory for those who grew up on the holiday coast, Deeside and along the A41 between Chester and Whitchurch! Add to these fortunate few the inhabitants of Buckley and the industrial villages west of Wrexham, who from their elevated positions, although more distant, possibly saw more of these airships as they floated above layers of low cloud and estuarine mist which allowed the vessels to creep up on town and village unawares and almost unnoticed. Memories zero in on R100, R101 and the Graf Zeppelin in particular.

R100, privately funded and erected by the Airship Guarantee Company at Howden (Yorkshire), was undoubtedly the finest airship ever built in Britain; R101, financed by the government and built by the Royal Airship Works at Cardington, was possibly the worst. Both represented the final attempt in this country to produce civilian airships for passenger carrying. The post-War rigid airship programme, which promised so much with R34, had ground to a halt with the disastrous structural failure of R38 in August 1921. Strikes and industrial unrest notwithstanding, both airships flew for the first time late in 1929. R100 made double crossings of the Atlantic to Canada in July and August 1930 and her commercial viability seemed assured.

Much to the surprise and delight of the North Wales coastal towns the R101 passed over them during an extended shake-down cruise over the week-end 18 November 1929 — a surprise because she was not expected. The itinerary as printed in the local papers was a post-flight handout, as was the accompanying self-congratulatory blurb which contains not a hint of the reservations and misgivings on the part of crew members. Delighted because the airship's progress was slow and ponderous and at low altitude — not for the benefit of spectators, but because of the vessels inadequate performance and lift., hence her withdrawal for extensions in length and cubic capacity — but all to no avail. Prestatyn UDC and Chamber of Trade could only bemoan the fact that R101's appearance was a week too early, that it would have done wonders for their 'Trade Week' beginning on 25 November. As the *Prestatyn Weekly* briefly commented under a headline DISTINGUISHED VISITOR: 'Many Prestatyn people caught a glimpse of that wonder of the air, the new airship, R101, on Monday morning when the magnificent vessel, the largest in the world, paid a very informal and unexpected visit ton the town. The air liner, flying at a low altitude and at an apparently slow speed, passed over the town at about 11.15 *a.m.* on its way home after the longest flight of its career, a distance of 1,800 miles in just over 30 hours. Admiration for the beautiful silver monster was experienced by all who were fortunate enough to perceive it, while the local Chamber of Trade confessed regret that the airship had not postponed its appearance before a Prestatyn audience until Trade Week when a civic welcome could have been accorded this giant of the clouds'.

Fullest details of R101's passage along the North Wales coast were, however, given by the *Flintshire Observer* after some frantic collation of numerous telephoned reports from their reporters who 'just happened' to be stationed along the R101's flight path!

The R100, seen here on the Cardington mooring mast, included North Wales on her shake-down itinerary 17/18 November 1929. [Ken Davies]

GIANT AIRSHIP OVER NORTH WALES.
R101's HIGHLY SUCCESSFUL FLIGHT.
SPLENDID VIEW OBTAINED AT LLANDUDNO

'A highly successful test flight by the world's largest airship, R101, was concluded on Monday evening when she returned to the mooring tower at Cardington after a run of about 1,800 miles in 30¹/2 hours. It was by far the longest trip of her short career and took her as far north as the Clyde and west to Belfast and Dublin. After she had been moored the officer in charge of the test expressed complete satisfaction with the flight which concluded the acceptance tests. All the men were able to get some sleep and a hot dinner was served in the saloon at midday. 'This is the timetable of R101's progress:

The R101 *was the 'government ship', built by and seen here at the mast of the Royal Airship Works at Cardington c1929.*

a.m.

　Left Cardington, Sunday 10.35

pm

　West of Hull 2.00
　Howden airship shed 2.30
　Newcastle 4.30
　Amble 5.20
　Edinburgh 7.00
　Glasgow 8.00
　Largs (Ayrshire) 8.50
　Isle of Arran 9.20

a.m.

　Isle of Man 12.45
　Blackpool 4.00
　The Skerries 7.00
　Dublin 8.15
　Anglesey 10.30
　Llandudno 10.50
　Rhyl 11.10
　West Kirby 11.20
　Chester 11.45

Over North Wales pm
　Lichfield 1.00
　Arrived Cardington 4.25

As a crew member revealed later 'We had an interesting race with the Holyhead Steam Packet boat. We gave it a quarter of an hour start from Dublin and after a few minutes overhauled it and reached Holyhead a full two hours before the steamer'. The chronological narrative resumes: 'The airship, coming from Ireland, passed over North Wales on Monday morning, flying very low and travelling slowly. School children and sight-seers were out and cheering and the men on the airship could be seen waving. The R101 passed over Holyhead at 10. 20 on Monday morning and was seen by many of the inhabitants. The visibility was good and she was moving majestically and slowly at a low altitude in a south-westerly direction. It was possible to discern her distinguishing numbers with the naked eye. She flew very low over Holyhead Mountain in the vicinity of which she turned almost about as if seeking her bearings and then headed for

Trearddur Bay. This slight diversion was probably made at the behest of R101's navigating officer, S/Ldr. E. J. Johnston, whose mother lived just outside Rhosneigr!

'Very few people in Bangor were aware that the R101 was passing over the outskirts of the city. She arrived from Holyhead and passed over the Menai Straits beyond Port Penrhyn. To all appearances she turned inland by Conwy, but reports show that she was seen at Llandudno and Colwyn Bay. Her advent near Bangor was hailed by the siren of a steamer, and this gave the impression that a fire had broken out — the Borough Fire Brigade are called out by this method! These who were fortunate to see the airship were impressed by her graceful movement, and of course, her size.

'The ship passed over Penmaenmawr and much interest was taken in its progress. Most of the quarrymen and residents had an excellent view of the airship as it passed over the quarry piers. Owing to the slow speed of the giant a splendid view was obtained. The airship passed over the Conwy Morfa travelling extremely low and at a slow speed. She did not pass over the town, but several people had an excellent side view. The airship made off towards Deganwy and Llandudno. The passage of the R101 over Llandudno which occurred at 11 o'clock on Monday morning will be recalled by all those who witnessed the spectacle. It flew low and apparently slowly and, passing as it did well over the heart of the town, it could not have been better staged had it been done deliberately for exhibition purposes. The big silver ship came gliding from the west glistening in the keen morning air. Every line was clearly seen, the black numbers on its side, the swinging gondolas which seemed strangely small in comparison with the huge bulk of the ship. In the town the people came running into the road from shops and houses to stare upwards as the round nose became visible over the roof tops as it approached from the direction of the West Shore. It followed a course almost parallel to Trinity Street, passed over the Police Station and on towards the promenade and the Little Orme. A thin trail of exhaust smoke marked its progress, and the low vibrant hum of its engines could be distinctly heard. It headed towards Colwyn Bay where some persons were notified by telephone of its approach. Considerable excitement was caused in Colwyn Bay when the airship was seen overhead. Visibility was good so that everybody had an excellent view of the silver giant which appeared to be largely skirting the coast in the direction of Liverpool. Actually it was on its way to Lichfield.

At Abergele many of the townsfolk had an excellent view of the airship which passed over the district at low level. The airship passed over Rhyl at about 11.10 am and was seen by a large number of people. It came from the direction of Llandudno and passed on towards Chester and the south-east.'

At every village school on the lower slopes of the Clwydian Range, lessons were halted and excited children ushered into the yard to view the passage of R101. The *Chester Chronicle* takes up the story:

'The passage of the R101 over North Wales was as dramatic as it was fleeting. At 11 o'clock it appeared as if the whole town of Llandudno had downed tools and crowded the promenade, squares and other vantage points. Children were permitted to bolt from school to catch a glimpse of the R101. For 15 minutes they were able to discern every feature of the passing airship until she glided eastwards to disappear into the mist. At 11.10 she was over Rhyl town and golf links, travelling south-east. R101 passed over Queensferry at about 11.30 and skirted West Kirby at 11.20. At a low altitude she passed over Parkgate a few minutes later, moving in the direction of Shotwick and Chester. At Chester spectators had taken up their positions on the city wall. At exactly 11.45 she came over the city from the direction of Parkgate. She was flying low — 1,000–1,500ft — and seemed to rise from behind some tall houses at the top of Canal Street. She made a circle around the cathedral tower. For a few minutes it looked as if she was heading for Liverpool. However, when in the neighbourhood of Hoole the airship doubled back over King Charles's Tower and was again headed for the centre of the city. Over Queen's Park she turned towards Whitchurch after manoeuvring over Chester for a quarter of an hour.'

Although people along the North Wales coast were taken largely unawares it would seem that no one had the nous to grab a camera and record for posterity the passage of this ill-fated airship. A *Chronicle* staff photographer didn't even have time to get into the street; he just leant through the office window, pointed his camera, pressed the button, and hoped for the best — a dark cigar-shaped silhouette amidst a swirl of fog and mist. As the airship's captain later commented, 'Chester was the last place we could see clearly and turning to the north-east we could see the great big pile of fog that was Manchester'

— hence the decision not to overfly that city. He continued, 'We struck fog around the Dee and by the time we reached Lichfield we had lost sight of the ground altogether. Occasionally we caught sight of it, just sufficient to correct our drift'. They saw the top 100ft of the Rugby wireless masts standing proud through the ground fog, but the Daventry masts were obscured altogether. Between Lichfield and Cardington they relied heavily on their D/F radio beams.

But from the outset R101 was beset by design problems that, because of political pressures, were never satisfactorily rectified before she set out on her ill-fated voyage to India on 5 October 1930. She crashed into a hillside near Beauvais, Department Oise, and caught fire instantly with 48 of the 54 people on board being burnt to death, including the captain, Major G. H. Scott, Llangefni's first CO, and S/Ldr. E. J. Johnston, who as a very green FSL also flew blimps out of Anglesey. The tragedy marked the end of Britain's involvement with airships and the R100 was scrapped. Some of the woodwork from the passenger accommodation found its way into the 'Crown & Kettle' Hotel, Great Ancoats Street, Manchester, a pitiful reminder of Sunday, 25 May 1930 when R100 passed over the city disrupting Rogationtide services.

With the loss of R101 and the scrapping of R100 a whole generation would elapse before a British airship again took to the air. It would be left to the German *Graf Zeppelin* to fill the gap in North Wales skies. She first flew in 1928 and, until broken up for scrap value in 1940, carried passengers in safety and comfort for more than a million miles, including over a hundred trans-Atlantic crossings and a world circum-navigation, all without loss of life. Although most of the city's inhabitants were sitting down to Sunday lunch and therefore missed the spectacle, the editor of the *Chester Chronicle* considered it an honour that her captain should take a SUNDAY MORNING PEEP AT CHESTER — his headline! It was a surprise visit as the city was not on the *Graf Zeppelin's* published itinerary, but the diversion may have had something to do with the fact that amongst its distinguished passengers was Lord Leverhulme and Col. Moore-Brabham, M.P. for Wallasey.

The German airship was making a 24-hour PR cruise around Britain over the week end Saturday/Sunday, 2/3 July 1932. She had arrived at Hanworth Park aerodrome (otherwise Feltham Aerodrome), Middlesex on the Saturday night to embark passengers and would fly

The Graf Zeppelin. *It was only in clear sunlight that the beautiful lines of the Graf Zeppelin could be fully appreciated. Its tour of North Wales in July 1932 was achieved in poor visibility.*

Despite thick fog for most of the time the Graf Zeppelin *attracted huge crowds when she visited North Wales and Chester in July 1932. A dark cigar-shaped shadow was all that was visible to the lens of the amateur cameraman as the airship hovered over the 'Carden Arms', Bulkeley, Cheshire.*

via Portsmouth (a new municipal aerodrome opened literally a few hours previously), north to Stranraer and Mull of Galloway, across to the Isle of Man, thence to Liverpool, across the Midlands to Birmingham and back to Hanworth via Cheltenham, Bristol and Southampton.

As the *Chronicle* smugly informed its readers, 'The *Graf Zeppelin*, following the example of the ill-fated R101 and the now dismantled R100, took a peep at Chester on Sunday morning… The noise first attracted attention despite the fact that not all engines were in action. There she was, shaped like an elongated bullet, beautiful in her proportions and painted a dull lead colour, The letters *GRAF* were plainly to be seen on her snub nose and nearby there was a lighter shade denoting a patch, or rather a mend in the outer fabric… The Zeppelin travelled at walking pace (or so it seemed). It was very slow to all appearances yet in fact her speed was probably 25 m.p.h. … She arrived from Liverpool over Chester College and cruised over the Town Hall Square and made for the Whitchurch direction over Beeston Castle. She was flying low, visibility was none too good and rain fell gently as she passed over the city'. Thus, rather poetically for a local newspaper hack, the curtain fell on the 'Golden Age' of airships as far as the north-west and North Wales were concerned. For the Midlands, too — it would be 40 years almost to the day before an airship next graced its air-space, and then, on 5 July 1972, only the Goodyear blimp *Europa* on one of its regular publicity and advertising sorties.

In April 1931, in an airship interstitial as it were, the attention of the whole country focussed momentarily on Bury Hill, Selhurst, Sussex, and on the north Shropshire village of Welshampton. In the latter parish 81-year old Preb. Henry Moody, rector for 45 years and Rural Dean of Ellesmere, was coming to terms with the sudden death of his only surviving twin son whilst on duty with the RAF. The first, 2nd/Lt. Charles Angelo Moody, RFC, only 18 years of age, had been killed in action over Belgium in the autumn of 1917. At 18.20 hrs. on 21 August 1917, in a Nieuport 23 (serial B1613) of 1 Squadron, he had taken off from Bailleul (Asylum Ground) on a 'Deep Offensive Patrol' Polygon Wood–Menin–Lille–Perenchies. Low air strafing and ground attacks were now a necessary prelude to assaults by British infantry. As

well as enemy strong points and communications, troops in billets behind the immediate battle area and targets around German HQ towns were attacked. 2/Lt. Moody was last seen at 19.10 hrs. in combat with twelve enemy aircraft over Houlthust. He was shot down by Ltn. F. Loerzer of *Jagdstaffel 26* at 19.15hrs.

His twin, Lt. (acting-Capt.) Henry Michael Moody, survived the war and elected to remain in the RAF, seeing service in India and the UK. He had joined the RFC as a mechanic in December 1916, trained as a pilot and was commissioned in June 1917. He was posted to 'A' Flight, 45 Squadron at Ste-Marie-Cappel, north of Hazebrouck, flying many sorties and claiming four enemy aircraft., one of which was the armoured Junkers J.l, the world's first all-metal plane designed especially for the close support of ground troops. On 21 September 1917 he had a narrow squeak as his Sopwith Camel B2320 suffered engine failure on patrol and he was forced to land — fortunately in British lines!

The squadron moved to Italy in December 1917 and Lt. Moody would notch up another four 'kills' flying his beloved Sopwith Camels out of Istrana (near Treviso) before posting to Home Establishment in June 1918. His two 'scores on 31 December 1917 were 45 Squadron's first on this front. At 09.45 hrs. his first combat sent an Albatros D.V spinning out of control into the ground at Piave de Soligo. It had been attacking a French observation balloon. At 10.30 hrs. over Paderno he forced a second Albatros to land, where it was captured. Ltn. A. Thurm of *Jagdstaffel 31* was fatally wounded. In April 1918 Moody was awarded the M.C. 'for carrying out very successfully a number of low flying patrols, photographic reconnaissances and escorts and showing on all occasions a very fine spirit of dash and determination'.

The war-time achievements of Welshampton's only 'ace' are listed below. There should possibly have been nine victories. On 27 February 1918 Moody, flying Camel B7283 of 66 Squadron, tangled with an Aviatik D.I, the first and finest all-Austrian fighter of the war, with, like the Camel, two fixed and forward-firing synchronised machine guns placed above the instrument panel. It was last seen, pilot apparently collapsed, in a vertical dive over Vittorio. Because he had not actually seen it crash, Moody's claim was disallowed by Wing

1917	*Aircraft*	*Location*	*Enemy Aircraft*
4 Sep.	B6238 Camel	Comines	Unidentified 2-seater — out of control
1 Sep.	B6238 Camel	Westroosebeke	D.F.W. C.l — out of control
20 Sep.	B6238 Camel	Passchendaele	Unidentified 2-seater — destroyed in flames
13 Nov.	B6238 Camel	Comines	Junkers J.1 — out of control
16 Dec.	*to Italy*		
31 Dec	B6238 Camel	Piave de Soligo	Albatros D.V — out of control
31 Dec	B6238 Camel	Parderno	Albatros D.V — forced to land/ captured
1918			
11 Jan	B6383 Camel	Corbelone	Albatros D.III — destroyed
30 Jan	B4609 Camel	Susegana	Albatros D.III — destroyed

In March 1931 F/Lt. Moody had been posted to 24 (Communications) Squadron, Northolt, operating a miscellany of aircraft to transport Government, Air Ministry and RAF 'VIPs'. On Thursday, 23 April 1931, Moody was personal pilot to AVM Felton Vesey Holt, C.M.G., D.S.O., AOC Fighting Area, Air Defence of Great Britain, who had his HQ at nearby Uxbridge. With another Moth as escort, they had just taken off from RAF Tangmere in D.H.60M 'Gypsy' Moth (K1838) of 24 Squadron after inspecting Nos. 1 and 43 Squadrons which were based at Tangmere aerodrome. Nine Siskin IIIs of 43 Squadron were airborne in V-formation at 2,000ft under the command of S/Ldr. L. H. Slatter. The latter dipped his flight in the accustomed dive salute.

Unfortunately he saluted the escorting Moth, piloted by F/Lt. E. H. Bellairs, which had taken off first. Moody's Moth, at 1,500 ft., was unsighted by the diving formation and was clipped by Siskin J8893, piloted by Sgt. C. G. Wareham, last in formation on the left leg of the 'V'. Although sustaining damage to the port upper wing, J8893 managed to regain the airfield. But K1838 was turned onto its back and sent spinning earthwards. AVM Holt managed to bale out, but hit the ground before his parachute could be fully inflated. F/Lt. Moody died instantly, crushed in the wreckage of his machine. Air Commodore A. L. Godman, writing in *The Times*, stated: 'By the death of H. M. Moody the RAF loses one of its most promising young officers. A brilliant pilot, devoted worker, good rider and shot, he was destined to rise to high rank. During the last three years his genius for command found a congenial outlet in the Apprentice Squadron of the Electrical and Wireless School, and a host of airmen who came under him will mourn his loss'. This latter unit was No. 1 E&WS at Cranwell (Lincs.) which was the central school for training both air and ground radio operators as well as providing courses on other types of electrical equipment. It had an establishment of aircraft to give flying training for wireless operators.

His father died in Ellesmere Cottage Hospital in October of that year — some say he never got over the loss of his only surviving son. Indeed the Welshampton War Memorial was hi-jacked by the ageing cleric as a particular memorial to his aviator sons. Despite public protest Henry Moody designed the memorial himself and defrayed the cost of its erection. The inscription reads: 'To the Glory of God and in Memory of their Son, Charles Angelo Moody, Lt. RFC who was killed in Belgium 21 August 1917, aged 18 years' — and almost as an afterthought: 'And of the other Honoured Heroes connected with this Parish who also gave their Lives in the Great World War 1914–18'. The names of the fifteen 'others' are engraved on the sides of the debased Celtic cross that rises above the memorial's stepped plinth. The front of the latter carries a later three line inscription in memory of F/Lt. H. M. Moody 'killed while flying on duty with his Air Vice-marshal'. Only when seeing the names of the twin brothers in close juxtaposition is one reminded that 'Michael' and 'Angelo' were born on Michaelmas day 1898. As if he were at long last being re-united with his sons, Preb. Henry Moody's grave lies immediately alongside the War Memorial, its Celtic cross headstone replicating the one he had designed a decade earlier. Such a disposition serves only to underscore the deep, inexpressible, unassuaged loss and pain that the reverend gentleman must have felt at the loss of his two boys. While he was yet rector of Welshampton the wooden cross that had temporarily marked young Charles Angelo's grave in France had an honoured place inside the parish church. Seventy years on, a sacrifice forgotten as ageing memories dimmed, the cross had been removed into the keeping of surviving members of the Moody family.

Strangely RAF records and/or the published version of the same also note Gypsy Moth K1838 as being in collision with another Siskin III, J8883, of 5 FTS Sealand, on the same date, 23 April 1931. This has to be wrong. The wreck of K1838 was complete. Although bits and pieces were salvaged it was never rebuilt and was officially struck off 24 Squadron's charge on 4 August 1931.

The *Shrewsbury Chronicle*, too, was not slow in linking events on the broader aviation scene with parochial Shropshire. They made good copy, a refreshing change from the ~stultifying minutiae of county life. Jet-setter business men half a century before their time were Lt/Col. G. L. P. Henderson, M.C., A.F.C. and Major T. B. Lloyd, the latter the son of George Butler Lloyd, Shelton Hall, Shrewsbury's M.P. 1913-22. On 24/25 April 1919, in the former's privately owned Armstrong Whitworth two seater biplane fitted with a 160 h.p. Beardmore engine (obviously a FK8 type), both men had flown London–Paris–Lyons–Nice–Dijon–Paris–Hounslow, some 1,375 miles. It was purely a business trip, with no advanced organisation, no mechanics carried, and no outside supporting help. In fact they had rebuilt the machine themselves from scratch after purchase as surplus. The FK8 had been used on various fronts for contact patrols, artillery spotting, light bombing, photo-reconnaissance etc. and had a maximum speed of 104 m.p.h. and a tank capacity of some 3½hrs, which makes the feat of these early continental trail-blazers all the more remarkable. Apart from enforced landings due to darkness, fog, rain-storms over the Alpes Maritimes, and the need to refuel and check oil levels, the round trip was accomplished in the remarkable time total flying time of 22.05 hrs. The engine ran perfectly 'with not a spanner laid on it or adjustment made!'

Others were not so fortunate. In the same month a British plane, London bound from Cairo, hit a tree whilst taking off from Bracciano airfield (some 20 miles north of Rome) and crashed to earth in flames. Captain Cecil H. Darley, pilot, was burnt to death. His brother, Major C. Curtis Darley, was severely injured.

The former had flown with the RNAS for 3½ years, logging 70 night bombing raids over enemy territory, distinguishing himself particularly by blowing up the Zeebrugge lock gates in May 1918, adding a D.F.C. to his D.S.C. In May 1919 he had made a memorable flight Ramsgate-Madrid in a Handley Page 'Super' via an enforced landing at Pau due to bad weather in the Pyrenees. Both were the sons of Charles Edward Darley, Caynton Manor, Shropshire.

Charles Curtis Darley was first commissioned in the RFA in 1910, but was also keenly interested in aviation. He gained the Royal Aero Club's Aviator's Certificate No. 513 in 1913 and was seconded to the RFC the following year. He specialised in reconnaissance work, in 1915 providing the first trench maps made from aerial photographs. It was on 26 October 1915, flying a Vickers FB.5 (5462) of 11 Squadron over Cambrai, that he was in combat with with the German air 'ace' Max Immelmann and was shot down, the latter's fifth victim. Darley's petrol tank was set on fire and bullets pierced his right arm, smashing his thumb and tips of fingers. Lt. Slade, his observer, performed a rough amputation with his penknife.

Darley, controlling the plane with his left hand, made a forced landing behind enemy lines and became a prisoner of war until repatriated in 1917. He remained in the RAF, becoming Superintendent of Physical Training RAF and Commandant of the RAF School of P. T., Uxbridge (1922–25). After a stint in India as CO RAF Kohat (1929–32), he would re-emerge on the Shropshire aviation scene as opener (1936) and first Commandant (1936–7) of No. 10 Flying Training School, Tern Hill. He was invalided out of the RAF in September 1939 with the rank of Air Commodore after a flying accident in India, but was re-employed in a civilian capacity at the Air Ministry. He died 10 June 1962, aged 72.

It was not often that a local newspaper editor sensed that he might have a 'big' aviation story only to find that further investigation came up against a wall of silence, diplomatic or otherwise. The *County Herald* (Flintshire) for 3 August 1928 harks back under the headlines ATLANTIC FLIGHT ECHO/STRANGE FIND AT FLINT to a possibly ill-conceived and certainly ill-fated east-west trans-Atlantic crossing attempt by the Hon. Elsie Mackay and Capt. Walter George Raymond Hinchliffe. They disappeared without trace, 'their fate a mystery, as was that of many others gone before them'.

The honourable lady was the daughter of the then Viscount Inchape of Strathnaver, Sutherland. Capt. Hinchliffe was an ex-RNAS/RAF 'ace' who saw considerable action on the Western Front with 10/210 Squadron in the early part of 1918 before being injured. His CV reads like something out of *Boys' Own*. He served with the Royal Artillery 1914-16, but after learning to fly (RAC Cert. No. 3595, 21 September 1916) at Redcar, Cranwell and Freiston, he was commissioned as a Flight Sub Lieutenant in the RNAS. In January 1918, after a stint as instructor at Cranwell (HMS *Daedalus*) he was posted to 10 (Naval) Squadron at Teteghan, near Dunkerque. He soon made his mark:

3 February 1918 flying Sopwith Camel B2604 raided Rumbeke aerodrome and forced down a red-nosed Albatros C seen to crash into trees SW of airfield at 1515hrs.

26 February 1918 in Camel B7190 dropped four 16lb bombs on Abeele aerodrome.

7 March 1918 in Camel B7190 dropped four 16lb bombs on Zeebrugge.

10 March 1918 in Camel B7190 brought down a German 2-seater near Roulers at 1445hrs.

1 April 1918 RAF formed, 10 (N) Sqn. became 210 Sqn. and Hinchliffe Lieutenant (acting-Captain) RAF.

3 April 1918 in Camel B7190 out of Treizennes, shot down German 2-seater in flames near Roulers at 1130hrs.

16 May 1918 in Camel D3387 out of St-Omer, shot down Albatros C, 'crashed and burst into flames near Bailleul at 1145hrs.'

18 May 1918 in Camel D3387 shot down Albatros DV over Neuve-Eglise at 1045hrs.

19 May 1918 in Camel D3387 shot down German 2-seater NE of Armentieres at 1045hrs.

But on 3 June 1918 Hinchliffe's luck ran out and he was badly injured whilst flying Camel C62 on a night patrol out of Ste. Marie-Cappel, near Hazebrouck. Two versions of the incident exist, one drawn from his service record and the other from aircraft movement and maintenance records. The latter have C62 as crashing during an attempted take-off after a forced landing near 32 Kite Balloon Station; the port tyre came off, the aircraft swerved and overturned. The pilot was injured, losing an eye. C62 was transferred the same day to No. 1 Aeroplane Supply Depot at Marquise where it was rebuilt as F5945 by 25 June. Its reincarnation was as brief as its earlier existence was long — taken on charge at No. 2 Air Issues Section at St. Andre-aux-Bois

on 10 August, by 73 Squadron at La Bellevue on 11 August, and was last seen spinning out of control over Marquion at 1030 hrs. on 5 September.

Hinchliffe's personal record explains the loss of an eye slightly differently. 210 Squadron was flying night patrols in an attempt to intercept the German Gotha G. V heavy bombers. The night was very dark, no moon, and a slight ground mist. Hinchliffe attacked one of the giant machines that had been picked out by searchlights over Hazebrouck. During this attack he may have been too confident and was well and truly caught by the Gotha's 'sting in the tail', a rear gunner who, by means of a wide ventral 'tunnel' could shoot both downwards and towards the rear, much to the discomfiture of unwary attackers. Hinchliffe was shot through the forehead, as a result of which he crashed on top of a forest, Nieppe Foret Dickebusch Lake, at 0115hrs. The machine overturned and was badly damaged, the pilot suffering severe facial injuries and losing the sight of his left eye. He wore a patch over this for the rest of his life.

Hinchliffe must have cut quite a 'romantic', dashing figure. A well-earned DFC was gazetted on 1 January 1919. His experience was considerable having flown over forty different types of aircraft - Avro 504, BAT Bantam, Blériot XI, Bristol F2b, Caudron G. III, Curtiss JN-4, de Havilland DH.4s, 9s, 16 and 34, Fokker F. 3, Handley Page 0/100, Nieuport Scout, Pemberton-Billing PB. 25, RAF BE. 2c and SE.5a, Sopwith Camel, Dolphin, Pup and 1½ Strutter, and the Vickers Vimy and Vulcan - to name but a few. Unlike so many of his contemporaries, he would appear to have successfully made the transition to civil aviation, holding Air Ministry's 'B' pilot's licence No. 235. He crops up in the records as a pioneer of pan-European routes first with North Sea Aerial & General Transport and then as chief pilot for KLM, the Dutch airline. . On 19 May 1921 he completed in a single day an Amsterdam-Dortmund-London-Amsterdam flight, total flying time 11¾hrs. On 22/23 July 1921 in a DH.9 he made a night-flight from Amsterdam to Berlin with passenger. The following day he completed a Berlin–Amsterdam–Lowestoft–London–Lympne flight in 12 hrs. 5 mins. flying time, before completing the circuit that same night in the first civilian Lympne-Amsterdam night flight with passenger. In 1922–3 Hinchliffe acted successively as chief pilot, Paris manager and Dutch manager to Daimler Airways before joining Instone Airlines with whom he flew all over Denmark, Germany, Holland and Belgium as well as on the London-Paris service. On 23 October 1922, flying a bright red de Havilland DH.34 (G-EBBS), Hinchliffe inaugurated the Daimler Airways Manchester-Amsterdam service via Croydon and Rotterdam with connections to Berlin. On 30 April 1923 G-EBBS would make the first through journey Manchester–Berlin. By 17 March 1924 the veteran pilot had logged some 5037 flying hours.

When on 4 May 1926, during the General Strike, the Prince of Wales rushed back to London from Biarritz, the Handley Page City of Pretoria was put at his disposal at le Bourget. A Vickers Vulcan, piloted by Captain Hinchliffe, now with Imperial Airways, was escort and 'back-up' plane, 'ready (according to the *Daily Mail*) to pick up the Prince and carry him along to his destination (Croydon) in case the first machine was compelled to make a forced landing'. On 20 December 1926 Hinchliffe flew the second Hercules to fly into Cairo, opening up Imperial Airways route to the Far East.

1927 had been a good year for Atlantic crossings and on 12 &3 April 1928 a Junker W.33 would make the first successful east-west crossing by an aeroplane, from Baldonnel, Eire, to Greenly Island, Labrador. If things had gone according to plan Hinchliffe had would have accomplished this feat a month earlier, in company with the Hon. Elsie Mackay, London stage and cinema actress, aged 34. In the light of what happened, some newspapers sought to dig up some juicy scandal between pilot and passenger, but failed. Little comment was made when Cmdr W. Stultz and Miss Amelia Earhart crossed the Atlantic on 17 April in a Fokker seaplane! But they, perhaps more sensibly, crossed west to east. On the evening of 12 March 1928, after some two weeks of careful preparation at Cranwell, Capt. Hinchliffe and the Hon. Elsie dined quietly at the 'George Hotel', Grantham, with an old acquaintance and fellow pilot, Capt. Gordon Sinclair. The pair left Cranwell at 0835 hrs. on 13 March, flying a black Stinson Detroiter monoplane with wings tipped in gold. Named the *Endeavour* it was powered by a 200 h.p. Wright 'Whirlwind' engine. At 1330 hrs. the chief lighthouse keeper at Mizenhead, Co. Cork, saw the Stinson over the village of Crookhaven, heading westwards out into the Atlantic, apparently on a direct course for Newfoundland. But they had already covered some 400 miles and had only just left Europe behind them. With hindsight a Lincolnshire aerodrome may not seem the most advantageous starting point for a trans-Atlantic crossing, a point not lost on Kohl, von Hunefeld, and Fitzmaurice in their successful attempt on 12 April.

Nothing further was heard from Hinchliffe and Mackay and they were duly reported missing. Great play was made over a psychic message allegedly received on 31 March 1928 by an elderly lady in Surrey: 'I DROWNED WITH ELSIE MACKAY. FOG, STORMS, WIND. WENT DOWN FROM GREAT HEIGHT OFF LEEWARD ISLANDS'.

Then on Tuesday, 31 July 1928, one George Dean of Marsh Lane, Flint, made a strange find on a little frequented stretch of the Dee estuary foreshore between Flint and Bagillt at a spot, as the Herald editor was quick to inform readers, 'in a direct line with the point at which the Dee enters the Irish Sea'. The find was a small square smelling salts bottle, once watertight, but now with top slightly damaged. Inside, much discoloured, was a tightly rolled slip of paper torn from a pocket book but with the words still discernible: 'Good-bye all. Elsie Mackay and Capt. Hinchliffe. Down in fog and storm'. It was handed in to Flint police, but in official circles 'little importance was attached to the find and it was dismissed as a cruel joke'. A puzzled editor concludes by asking, and receiving no convincing reply, why the handwriting on the note had not been submitted for identification by either Lord Inchape or Miss Hinchliffe. Seventy years on a new generation of researchers still strives for an answer, but the details, as revealed in a local Flintshire newspaper with restricted circulation, might be the last piece of the jig-saw to be put in place.

Bibliography

P. Abbott, *Airships* (Aylesbury, 1991)

P. Abbott, *The British Airship at War, 1914-1918* (Lavenham, 1989)

F. J. Adkin, *RAF Ground Support Equipment since 1918* (Shrewsbury, 1996)

E. Angelucci, P. Matricardi, *World Aircraft: Origins—World War 1* (Maidenhead, 1977)

C. Ashworth, *Action Stations: Military Airfields of the South-West* (Wellingborough, 1982, 1990)

J. A. Baker, N. Pritchard, *Balloons and Ballooning* (Aylesbury, 1986)

M. J. F. Bowyer, *Action Stations: Military Airfields of Oxfordshire* (Wellingborough, 1988)

M. J. F. Bowyer, *Action Stations: Military Airfields in the Cotswolds and the Central Midlands* (Wellingborough, 1983, 1990)

A. Brew, *A History of Black Country Aviation* (Stroud, 1993)

J. M. Bruce, *The Aeroplanes of the Royal Flying Corps (Military Wing)* (London, 1982)

P. H. Butler, *British Isles Airfield Guide* (Liverpool, 1965, 1973, 1981)

F. J. Camm, *The Flying Reference Book* (London, 1939)

W. J. Claxton, *The Mastery of the Air*, Edinburgh, 1915.

K. Delve, *The Source Book of the RAF* (Shrewsbury, 1994)

H. Driver, *The Birth of Military Aviation: Britain 1903-1914* (Woodbridge, 1997)

G. Endres, *British Aircraft Manufacturers since 1908* (Shepperton, 1995)

A. P. Ferguson, *A History of RAF Shawbury* (Liverpool, 1977)

A. P. Ferguson, *A History of RAF Sealand* (Liverpool, 1978)

P. Francis, *British Military Airfield Architecture* (Yeovil, 1996)

N. Franks, R. Guest, G. Alegi, *Above the War Fronts* (London, 1997)

C. Grahame White, *Aviation,* (London, 1912)

Grahame White Company Souvenir, (1917)

J. J. Halley, *The Squadrons of the RAF & Commonwealth 1918-1988* (Tonbridge, 1988)

B. B. Halpenny. *Action Stations Military Airfields of Greater London* (Yeovil, 1984, 1993)

D. W. Harris, *Maritime History of Rhyl and Rhuddlan* (Prestatyn, 1991).

T. Henshaw, *The Sky their Battlefield* (London, 1995)

R. Higham, *Bases of Air Strategy: Building Airfields for the RAF, 1914–1945,* (Shrewsbury, 1998).

R. Hough, *The Great War at Sea 1914-1918* (Oxford, 1983)

R. Jackson, *Airships* (London, 1971)

Jane's Fighting Aircraft of World War I (London, 1993)

C. G. Jefford, *RAF Squadrons* (Shrewsbury, 1988)

P. Kemp, *U-Boats Destroyed: German Submarine Losses in the World Wars* (London, 1997)

B. King, *Royal Naval Air Service 1912-1918* (Aldershot, 1997)

P. King, *Knights of the Air* (London, 1989)

A. McKinty, *The Father of British Airships: A Biography of E. T. Willows* (London, 1972)

L. Marriot, *British Military Airfields Then & Now* (Shepperton, 1997)

F. K. Mason, *The British Bomber since 1914* (London, 1994)

F. K. Mason, *The British Fighter since 1912* (London, 1992)

J. H. Morrow, *The Great War in the Air: Military Aviation from 1909 to 1921* (Shrewsbury, 1993)

C. Mowthorpe, *Battlebags* (Stroud, 1995)

G. Negus, T. Staddon, *Aviation in Birmingham* (Leicester, 1984)

R. C. Nesbit, *The RAF in Camera, 1903-1939* (Stroud, 1997)

D. Oliver, *Hendon Aerodrome, (*Shrewsbury, 1994).

D. Peel, *British Civil Aircraft Registers since 1919* (Leicester, 1985)

J. Penny, *Robert Cadman, Steeple Flyer* (Bristol, 1989)

H. Penrose, *British Aviation: The Pioneer Years* (London, 1967)

B. Quarrie, *Action Stations: 10. Supplement and Index* (Wellingborough, 1987)

J. D. R. Rawlings, *Coastal, Support and Special Squadrons of the RAF and their Aircraft* (London, 1982)

J. D. R. Rawlings, *Fighter Squadrons of the RAF and their Aircraft* (London, 1993)

B. Robertson, *British Military Aircraft Serials 1878-1987* (Leicester, 1987)

B. Robinson, *Aviation in Manchester* (Manchester, 1977)

L. T. C. Rolt, *The Aeronauts* (London, 1985)

S. W. Roskill (ed.), *Documents Relating to the Naval Air Service* (London, 1969)

RAF Tern Hill, A Brief History of (Tern Hill, 1975)

C. Shores, N. Franks, R. Guest, *Above the Trenches* (London, 1990)

C. Shores, C. Williams, *Aces High* (London, 1994)

R. Sloan, *Early Aviation in North Wales* (Llanrwst, 1989)

D. J. Smith, *Action Stations: Military Airfields of Wales and the North-West* (Wellingborough, 1981)

R. Sturtivant, J. Hamlin, J. J. Halley, *Royal Air Force Flying Training and Support Units* (Tunbridge Wells, 1997)

R. Sturtivant, G. Page, *The Camel File* (Tonbridge, 1993)

R. Sturtivant, G. Page, *Royal Navy Aircraft, Serials and Units 1911-1919* (Tonbridge, 1992)

R. Sturtivant, G. Page, *The S. E. 5 File* (Tonbridge, 1996)

J. C. Temple, *Industrial Archaeology of Aviation in Shropshire* (unpublished thesis, Ironbridge Institute, 1984)

A. J. Tennent, *British Merchant Ships sunk by U-boats 1914-19 (* 1990)

O. Thetford, *Aircraft of the Royal Air Force since 1918* (London, 1995)

O. Thetford, *British Naval Aircraft since 1912* (London, 19171)

D. Thompson, R. Sturtivant, *Royal Air Force Aircraft J1-J9999 and WW1 Survivors* (Tonbridge 1987)

D. W. Williams, *Heroic Circumstances* (Ruthin, 1997)

T. B. Williams, *Airship Pilot No. 28* (London, 1974)

K. Wixey, *Gloucestershire Aviation: A History* (Stroud, 1995)

P. Wright, *The Royal Flying Corps 1912-1918 in Oxfordshire*